Thank you Marsha
for all you do for
us and for snow leopards!

Chan

The Partners Principles for Community-Based Conservation

Charudutt Mishra, Ph.D.
Science and Conservation Director Snow Leopard Trust & Trustee,
Nature Conservation Foundation

Published by:
Snow Leopard Trust www.snowleopard.org

First published 2016

Printed on recycled paper at:
Multiplexus (India)
www.multiplexusindia.com

First Edition

ISBN: 978-0-9773753-1-8

ISBN: 978-0-9973753-0-1 (e-book)

Mishra, Charudutt (2016) The Partners Principles for community-based conservation. Snow Leopard Trust, Seattle, USA.

Contact info@snowleopard.org for further information

Cover photo by Paul Sangeorzan

Contributors

Idea:

Fred Nelson, Acacia Conservation Fund

Copy-Editor:

Matthias Fiechter, Snow Leopard Trust

Project Editors:

Brad Rutherford, Snow Leopard Trust

Matthias Fiechter, Snow Leopard Trust

Chapter Co-Authors:

Agvaantseren Bayarjargal (Chapters 10, 11), Snow Leopard Conservation Foundation & Snow Leopard Trust

Ajay Bijoor (Chapter 11), Nature Conservation Foundation & Snow Leopard Trust

Brad Rutherford (Chapters 10, 11), Snow Leopard Trust

Gina Cantara (Chapter 10), Snow Leopard Trust

Jennifer Snell Rullman (Chapter 10), Snow Leopard Trust

Karma Sonam (Chapter 11), Nature Conservation Foundation & Snow Leopard Trust

Kulbhushan Singh R. Suryawanshi (Chapters 10, 11), Nature Conservation Foundation & Snow Leopard Trust

Muhammad Ali Nawaz (Chapter 12), Quaid-i-Azam University, Snow Leopard Foundation, & Snow Leopard Trust

Radhika Timbadia (Chapter 11), Nature Conservation Foundation & Snow Leopard Trust

Tanzin Thinley (Chapter 11), Nature Conservation Foundation & Snow Leopard Trust

Review and Inputs:

Aparajita Datta, Nature Conservation Foundation

Brad Rutherford, Snow Leopard Trust

Juliette Young, Centre for Ecology and Hydrology, Natural Environment Research Council

Justine Shanti Alexander, Snow Leopard Trust

Mahesh Rangarajan, Ashoka University

Rohan Arthur, Nature Conservation Foundation

Stephen R. Redpath, Aberdeen University

Yash Veer Bhatnagar, Nature Conservation Foundation & Snow Leopard Trust

Indexing and Logistics:

Rakhee Karumbaya, Nature Conservation Foundation & Snow Leopard Trust

In memory of Lkhagvasumberel Tomorsukh (1988 – 2015)

You live on in our thoughts and in our work, motivating us to do more and inspiring us to do better. Rest in peace, dear Sumbee.

Table of Contents

Preface

The original idea for this book came from Fred Nelson of the Acacia Conservation Fund. Fred encouraged us to write about our community-based programs to conserve snow leopards *Panthera uncia* and their habitats in the mountains of Asia. The Snow Leopard Trust and its partners have engaged with local communities and national governments in key snow leopard habitats of Central Asia for nearly three decades.

Brad Rutherford of the Snow Leopard Trust thought Fred's was a great idea, and I fully agreed. Fred followed up by putting money where his mouth is. His idea, and a grant from the Acacia Conservation Fund, made this book possible.

I remember completing the first 5000 words or so and sharing them with Brad. I sent him the document over email, that communication miracle which didn't exist for half of my life, but is rapidly making up for its absence by occupying most of my time in the second half.

A discussion followed over the other communication miracle, called voice-over-IP, or simply Skype for the uninitiated like me. To save on bandwidth, we did not have our cameras on (do they call it video-over-IP?), so I am still uncertain whether Brad was utterly delighted or utterly dismayed by what he had read. It seemed to be one of the two, and I thought it best to not probe further.

Looking for practical solutions comes naturally to Brad. I, on the other hand, was struggling, trying to articulate some practical solutions for community-based conservation. That Skype discussion drove home the point that writing a practical guide to community-based conservation was not going to be an easy task.

Like any other activity that so closely involves people, community-based conservation is a complex pursuit. Combine that with the complexity of natural systems, and you have an almost impossible task at hand, especially when it comes to writing guidelines. No questions have simple answers. Every issue is nuanced in its ecological and human complexity. There isn't and perhaps shouldn't be any single way to do community-based conservation.

The purpose of this document is to share our experiences and thinking, and to codify an approach for conservationists to work effectively and respectfully with local communities. I have distilled this approach into eight broad principles, many of which will resonate not just with conservationists, but others involved in human development and community work. My aim is to help others understand why some of our efforts appear to have worked. To help them not make the mistakes we made.

I must thank Brad for his coaxing, patience and tolerance. This isn't perhaps what he was initially expecting, but he still allowed and encouraged me to pursue it. I remain grateful.

I am forever indebted to the countless people from mountain communities who have welcomed me to their homes in the Asian highlands, ranging from Eastern Himalayas and Western Trans-Himalayas in south Asia, to the Afghan Pamirs, the Hindukush, and the Kyrgyz Tien Shan in the west, the Mongolian Altai in the north, and the Kunlun Shan and the Tibet-Qinghai plateau right in the heart of the continent.

These wonderful people have hosted me, and made the time to interact with me despite their busy and challenging lives. They have gone out of their way to help me not to feel like an intruder, guided me, and have enriched my life in many ways. They are too many to name, though I must mention my long-time friends Karma Sonam, Tanzin Thinley, Lobsang Gyalson 'Sherpa', and Sushil Dorje from whom I have learnt much.

I am grateful to many friends and colleagues at the Nature Conservation Foundation. I especially thank Aparajita Datta, Rohan Arthur, M.D. Madhusudan, and Yash Veer Bhatnagar, who have been a significant part of my work. The actual work that forms the cornerstone of this book has been conducted by several staff based at Snow Leopard Trust and our range country partners including the Nature Conservation Foundation in India, Snow Leopard Foundation in Pakistan, Shanshui in China, Snow Leopard Conservation Foundation in Mongolia, and Snow Leopard Foundation in Kyrgyzstan. Siri Okamoto has been a brilliant enabler of all our work, and my conscience keeper too. Ever since I scribbled the first few pages, my editor Matthias Fiechter has been a constant and significant contributor to the development of this book. Deborah Turnbull and Chris Czarnecki have helped tremendously in minimizing typographical errors.

During my formative years, it was my mentor Herbert Prins who nudged and enabled me to start working with village communities. I have since benefited a lot from my interactions with friend and collaborator Steve Redpath, who readily agreed to write the foreword for this book. Juliette Young made this book actually worth something when she developed a training program based on the PARTNERS Principles for practicing conservationists, one that has been highly appreciated by our participant trainees from various countries. Ajay Bijoor has been working with communities and has also been effectively delivering the training program to conservationists.

The Whitley Fund for Nature, more recently with Fondation Segré, has been an incredible supporter of community-based conservation globally. Their unwavering support to me since 2005 has enabled a large part of my growth as a conservationist, and my engagement with local communities, which in turn led to the PARTNERS Principles and this book.

Charudutt Mishra
2016

Foreword

As I write this there are estimated to be 7,428,219,456 people living on earth, with over 2 billion of those living on less than $3.10 a day. We share our planet with an estimated 8.7 million species, and this diversity is coming under increasing threat from human activities. We are therefore faced with the dual, interwoven imperatives of helping improve the quality of life amongst our poorest people and conserving our biodiversity.

Our ability to achieve conservation objectives in this setting requires a sea-change in how conservationists traditionally think about and work with people, whether they be the rural poor, governments or businesses. Conservation will continue to struggle when it is at odds with the needs of those who live and work in nature and we need to stop thinking of people as the problem, but as part of the solution. Indeed, conservation is littered with examples of inappropriate interventions that have led to serious injustice to local people and have ultimately set back conservation efforts. In many cases such problems arise from the genuine concern that we are in the midst of a biodiversity crisis and so there is an urgent feeling that we need to act rapidly. However, it is far more important to act appropriately. We need to change from focusing primarily on the policies and top-down control of people's activities, to building shared solutions based on strong relationships and trust.

Many of us get into conservation because we care about the natural world and are interested in ecology. We want to go off and "save the world". But we have little training in how we should engage with communities and what various aspects we need to think about when we do. This is why this book is so incredibly valuable. It is a handbook that will help conservationists become aware of the pitfalls and find a path towards effective community-based engagement for conservation.

As the book points out, there is no one way to do community-based conservation, but this book will help guide your thinking and identify the important aspects for consideration. Charudutt Mishra and colleagues in the Snow Leopard Trust have done a wonderful job in distilling decades of experience of working with communities in central Asia. The primary focus of their work has been on snow leopard conservation, yet the principles for community engagement that have arisen from their experiences are applicable to a great diversity of conservation projects.

I was struck by two aspects of this remarkable book. Firstly at its heart there is a deep love of both nature and humanity. The excitement, the joys and the challenges of community-based conservation shine out from its pages. Secondly, there is a refreshing honesty and a recognition of the doubts, problems, confusions and failures, and, crucially, a willingness to embrace these aspects and learn from them. Conservation will always face different challenges and opportunities and we have to continue to learn and adapt and strive to improve our interventions with communities.

This book deserves to be widely read and I hope that it will encourage others working with communities, policy makers and businesses in different systems to analyse their successes and failures and share them with the rest of us. Time is indeed pressing and the sooner we can learn from these different examples the sooner we can hope to gain more success in conservation.

Professor Steve Redpath
University of Aberdeen.
June 2016

Part 1:
THE PARTNERS PRINCIPLES FOR COMMUNITY-BASED CONSERVATION

Chapter 1:
Introduction to the PARTNERS Principles

Conservation amidst people

Biodiversity conservation efforts aim at perpetuating the survival and functioning of wild species and ecosystems. Today, they form important – though often compromised – elements of national and global laws, policies, and conventions.

Biodiversity conservation usually involves trading off short-term and direct resource use and socio-economic benefits in exchange for more diffused, longer-term societal gain such as maintaining biodiversity, ecosystem services, and other economic, aesthetic, or spiritual benefits.

Over the last several decades, the establishment and management of Protected Areas by states has been the standard approach to biodiversity conservation globally. In practice, these efforts have generally entailed, and, indeed, expected, diminished resource access as well as reduced economic development of individuals in order to achieve larger societal conservation goals.

In large parts of the world, the main costs of conservation continue to be borne by the relatively poor, living in and around Protected Areas or generally important biodiversity areas. The cost of conservation to local communities due to curtailed access to natural resources, ecosystem services, and developmental programs are further aggravated by wildlife-caused damage, including injuries or loss of human life, and economic and psychological impacts (Madhusudan and Mishra 2003). Such damage results in retaliatory killing of wildlife and erosion of support for conservation efforts. It also leads to resentment among local people, in part due to the inability to take retaliatory measures without attracting punitive legal action.

Such costs lead to disenchantment among local communities and their political representatives regarding conservation efforts, and to serious disagreements with conservationists and managers. The resultant protracted conflicts have been generally referred to as human-wildlife conflicts, and more recently and perhaps more appropriately, conservation conflicts (Redpath et al. 2013). The lack of local community support for conservation is one of the most important factors undermining global efforts to protect biodiversity today.

Yet, the predominant measures to achieve biodiversity conservation continue to be state-imposed and focused on law enforcement, with little space for meaningful local community participation. Conservation, especially the management of Protected Areas, is therefore often perceived to be discriminatory by local people (Mishra et al. 2003a, Bhagwat and Rutte 2006).

On the one hand, conservation efforts have tended to be top down, coercive and perceptibly unjust. On the other hand, conservation is considered to have a relatively strong moral basis, and appeals to human value systems. The irony of real world conservation is hard to miss.

Community-based conservation: complicated but necessary

Hardin (1968) formalized the idea that degradation of natural resources was the predictable outcome of an increase in human population and per capita resource consumption, interacting with rational human self-interest. In a system where the individual derives a direct, immediate benefit from exploiting natural resources while suffering only delayed costs of collective over-exploitation, Hardin's model predicted the inevitability of degradation of natural resources, or the tragedy of the commons; "Ruin is the destination towards which all men rush, each pursuing his own best interest..." (Hardin 1968 p. 1244).

In Hardin's scheme, centralized government control or privatization of natural resources were the two institutional arrangements that could prevent the inevitable degradation of natural resources (Hardin 1968, Ostrom 1990, Dietz et al. 2003). Others also proposed related ideas, such as Olson in 1965, in the logic of collective action. Here too, rational self-interest was predicted to prevent individuals from achieving group interests unless the group was very small, or there was coercion (Olson 2009).

Notwithstanding the rationale – and the abundant examples – of the commons' tragedies, societies have also repeatedly and sometimes effectively responded to natural resource scarcity, and have developed self-governing systems of resource management. In contrast to the views of Olson and Hardin, they have done so without the need for coercion. They have done so without centralized governments or blanket privatization of natural resources; the only two institutional arrangements capable of sustaining the commons in Hardin's model.

Elinor Ostrom, an influential figure in the research program which developed in Hardin's wake – and which questioned the universality of the models of commons' tragedies – examined both functional and failed self-governance systems. She identified fundamental characteristics of common-pool management, and distilled them into a set of eight design principles that presumably influence the success of self-governing local institutions (Ostrom 1990). In Ostrom's view, tragedy of the commons was not the inevitable consequence of common-pool characteristics interacting with individual self-interest. On the contrary, the tragedies of the commons, common as they may be today, may be viewed as signs of unstable institutions.

Yet, the models of commons' tragedies have been highly influential. It is perhaps no mere coincidence, for example, that today's paradigm of biodiversity conservation is founded on the idea of Protected Areas, that governments currently manage the majority of existing Protected Areas, and that setting up Protected Areas has often involved the coercion of local people.

Where, then, is the space for community-based conservation? Community-based conservation approaches recognize the important role of local communities in biodiversity conservation. Through direct involvement and empowerment of local people in conservation and conflict management, and through indirect efforts such as helping them improve their quality of life, community-based efforts try to reduce the disproportionate burden of conservation costs that the local communities bear, and thereby seek their support for conservation. Community-based conservation efforts try to assist people in self-governance of natural resources and biodiversity. The importance of local community involvement in biodiversity conservation is increasingly emphasized in policies and environmental rhetoric.

Community-based conservation usually involves assisting local communities in maintaining or strengthening their conservation-friendly practices, changing their conservation-unfriendly practices or internal threats, and collaboratively addressing external threats to biodiversity. They also try to promote the ownership and accountability of natural resources among local communities.

At first glance then, the models of local self-governance of common-pool resources come across as useful frameworks for community-based conservation. And while they are indeed very useful, they may not be sufficient; today's reality of commons' tragedies cannot be ignored. The rate of depletion of biodiversity and natural resources has increased catastrophically, due to large-scale industrial exploitation, climate change, and commons' tragedies. While local human institutions that sustained natural resources over considerable periods of time still exist in places, the conditions for effective governance of common-pool resources are increasingly rare (Dietz et al. 2003).

It is therefore in this divergent, complicated space of human behavior and its correlates that the idea of community-based conservation must be conceptualized and explored. Even as theory develops around them, it is not one, but both of these influential bodies of work – the tragedy of the commons on the one hand and the self-governance of the commons on the other – that provide the conceptual foundations for examining and informing global efforts to conserve biodiversity. Indeed, Ostrom herself reported how self-governing institutions that were successful in resource management were '...rich mixtures of public and private instrumentalities' (Ostrom 1990). While such an academic exploration is not the purpose of the present book, recognizing these underlying conceptual dichotomies at the outset is a necessary acknowledgement of the complexity of conservation with communities.

This complexity isn't restricted to the underlying social science models. Ecosystems function in intricate, often non-linear ways. Human behavior, similarly, is highly variable across time and across individuals, as are human aspirations. Then there is the political ecology of accessing common-pool resources. The social institutions relevant in biodiversity conservation are diverse and operate at multiple scales, from the local community to provincial and national governments, to international organizations. Biodiversity conservation is thus a multi-level commons problem with complex issues of ownership and control (Berkes 2007). Community-based biodiversity conservation, even more so. These complexities surrounding biodiversity and natural resources, and the dynamism of human behavior and societies make community-based conservation a complicated undertaking.

However complex it might be, conservation practitioners must engage with communities. Not only because resource use by local communities impacts biodiversity, or because they can be very useful and influential partners in conservation. Nor just because an exclusive focus on Protected Areas is ecologically inadequate to conserve landscape species like the snow leopard. Conservationists – especially those who find themselves not able to morally reconcile with the unfairness of top-down conservation efforts – have little choice but to get involved in community-based conservation.

The need for community-based conservation frameworks

The management of Protected Areas worldwide has been formalized and standardized in terms of governance, categorization, and administration (e.g. Dudley 2008) to quite a significant extent. The approach is largely in line with the models of commons' tragedies. The situation with community-based conservation, whether in Protected Areas or otherwise, is different. Despite academic foundations being available – especially of self-governance of natural resources (e.g. Ostrom 1990) – there are no clear-cut frameworks or universally accepted guidelines for conserving with communities.

Because community-based conservation encompasses a diversity of scales, institutions, and perspectives, it is considered a complex systems problem, and therefore, pluralism in approach is both to be expected and valued (Berkes 2007). Recognition that there may be multiple pathways to achieving the same goal is an important aspect of working with communities. But this is perhaps not the only reason why we lack universal guidelines for community-based conservation.

An important motivation for community-based efforts is the acceptance that in democratic systems, conservation efforts are less impactful and difficult to sustain without the support of local people. Often, it is also the role of personal values, dilemmas, and empathy of the conservationist that propels individual practitioners towards community-based conservation. This has implied a considerable influence of individual interest, values and worldview of the practitioner on the approach and the interventions employed in community-based conservation efforts.

As conservation practitioners engaging with communities, we learn from experience, from trial and error. Important though that is, it doesn't take away the need for some practical and general guidelines or frameworks for community-based conservation – or at least for some resources that one could consult in order to learn from the experiences of others. To adopt, adapt, or, at the least, to consider their best practices. To avoid making the same mistakes they made.

While pluralism in community-based approaches is to be highly valued, the paucity of efforts to consolidate the learning and experiences in community engagement remains a constraint in conservation thought and practice.

This work

What this book is about

The present book is a response to this lacuna. It is aimed at sharing experiences in community-based conservation efforts focused around the snow leopard in Central Asia, but is expected to have wider relevance. It is an attempt to distil our experiences – together with insights from various disciplines such as ecology, sociology, social psychology and negotiation theory – into a set of principles that together constitute an approach to community-based conservation.

The Snow Leopard Trust and its partners have been involved in working with local communities in the Himalayas and Central Asia for many years to promote the conservation of the snow leopard and associated biodiversity. We use the term community to denote a hamlet or village, a collection of individuals or households who identify themselves as a community, live in the same area, and share systems of local resource use, traditions and governance.

We have been running several community-based conservation interventions, supported by education and awareness initiatives wherever possible. These have included a collaborative corral improvement initiative that, together with the local people, aims to reduce livestock losses to predators. When livestock depredation does occur, our community-based livestock insurance program aims at sharing and offsetting economic losses to local people (see Chapter 11: Livestock Insurance). We also run interventions that aim at improving the social carrying capacity for predators by linking livelihood enhancement to conservation action, or conservation friendly behavior. Our program Snow Leopard Enterprises (SLE) is an example of this approach (see Chapter 10: Snow Leopard Enterprises). SLE involves training local people to produce handicrafts that are marketed regionally or internationally. The livelihood opportunity is provided in exchange for community support in preventing poaching in their area. A built-in reward system, where the community is entitled to a bonus on all purchases if the conservation commitment is met, creates a positive incentive for wildlife protection. Another intervention to promote tolerance of predators, called the Ecosystem Health Program, involves a snow leopard friendly livestock vaccination program in areas where local communities do not have access to adequate veterinary healthcare (see Chapter 11: Livestock Vaccination). Our intervention portfolio has also included a system to establish informal village wildlife reserves (Mishra et al. 2016a). This involves collaboratively curtailing grazing and natural resource harvest from selected areas on local community land to enable wildlife recovery.

In the course of running these interventions, we have had many positive experiences. But we have also made mistakes. There are several things we would do differently if we did them again. While our experiences have been mixed, one thing has remained unchanged. We continue to remain strong advocates of local community involvement in conservation.

This book aims to articulate the approach that we believe should be followed when working with local communities. The approach is crystallized in the form of eight broad principles, which, for simplicity and retention, are acronymed the PARTNERS Principles for Community based Conservation. The acronym is more than a catchword. It underscores our deep conviction that local communities must be equal partners in conservation.

The second part of this book includes detailed descriptions of three of our community-based interventions, viz., Snow Leopard Enterprises, livestock vaccination program, and community-based livestock insurance program. These are written in a manner that would be useful to the conservation practitioner. Each of the initiatives is discussed in the context of the PARTNERS Principles.

The book is expected to be useful for the conservation practitioner involved in community-based efforts. It will also be pertinent for those practitioners who feel the need and the value of community involvement in conservation, but are unsure of how to proceed, or hesitate to do so for fear of making mistakes. This work might also be of use for grant-making organizations and professionals to consider some of the lessons we have learnt. And it might even interest those who do not believe in the value of community-based conservation, if nothing else, at least for taking the debate forward.

In some sense, this work has relevance to all who care for biodiversity conservation. Especially for those who would like our magnificent wildlife and biodiversity to be conserved, but for whom it is important that conservation is achieved in a just and equitable way – those for whom it is not just the end that matters, but also the means.

What this book isn't

Measuring the impact of community-based conservation efforts is an important issue that deserves much attention, but it's not the subject of this book. Several critical questions need to be answered. What are the impacts of community-based conservation action on biodiversity or focal species populations? What are the impacts on the threats they face? How do community-based conservation efforts influence peoples' conservation-related behavior? What are the larger social impacts of community-based conservation actions?

Perhaps an even more important question to tackle is: what are the correlates or determinants of performance of community-based conservation efforts? The models of self-governance of the commons, especially Ostrom's design principles, provide a highly useful conceptual space for such analyses. This book, however, does not attempt to tackle these critical questions, integrally related and much needed as they are. I do believe that a more in-depth and critical analysis of the PARTNERS Principles from the perspective of self-governance models would be highly insightful, and should be a subject for future work.

My goal here, however, is more humble: to provide a framework that distills some best practices for community-based conservation based on our experiences. Indeed, some of these best practices do have a critical influence on the performance of community-based conservation efforts. Others address issues of fairness and basic human dignity. The PARTNERS Principles are a blend of the practical and the ethical.

The PARTNERS Principles

Figure 1.1 A detailed visual representation of the eight PARTNERS Principles for effective and respectful community-based conservation.

The PARTNERS Principles underscore the critical importance of a set of 8 criteria for effective community-based conservation programs (Figure 1.1). In the following chapters, each of the principles is described in detail, with relevant examples. They include:

1) Relationship-building through the sustained and long-term Presence of conservationists amidst the local community (Chapter 2).
2) The Aptness of specific community-based interventions with respect to addressing the main threats to biodiversity, the underlying science, the

local culture, socio-economics, the available or potential social capital, and the value of multi-faceted programs (Chapter 3).

3) A relationship that views the community with dignity and Respect, and interactions based on beneficence and non-malfeasance (Chapter 4).
4) High Transparency in interactions with local communities with truthful and open communication regarding each other's interests, and visible equitability in program benefits to community members (Chapter 5).
5) Integrative Negotiations with local communities and interventions based on formal agreements and conservation linkages (Chapter 6).
6) The ability to view problems, constraints and opportunities from the community's perspective with a high level of Empathy (Chapter 7).
7) The ability to adaptively improve the programs and address emerging problems and opportunities with a high level of Responsiveness and creativity (Chapter 8).
8) Strategic support (Chapter 9) to increase the resilience and reach of community-based conservation efforts through partnerships with governments in management planning and implementation, and policy and legal support.

Chapter 2:
PRESENCE

Community-based conservation cannot be done effectively from a distance. It is founded on resilient relationships between local communities and conservationists, which require the practitioner's sustained field presence. Inadequate field presence and participation of conservation organizations is perhaps a larger constraint for effective community-based conservation, compared to the extent of participation of local communities.

While it is neither possible nor necessary to be present in each community, being based in a relatively larger community in the focal conservation landscape, with periodic visits to others, is useful in building strong relationships with local people, and generating current and contextual knowledge. Importantly, such immersion also serves as an early warning system to track and tackle new and emerging threats to biodiversity.

Being there

How does one engage with local communities? In project proposals submitted to funding agencies, and in practice, community engagement is typically suggested to be undertaken through periodic visits and structured workshops and meetings with local people. These are no doubt useful when conducted in an inclusive and democratic manner. They help to evolve and formalize conservation agreements with the community transparently, and to better monitor and evolve the programs adaptively. They may even be adequate for successful joint implementation of initiatives that benefit the community. However, by themselves, these are unlikely to achieve effective on-ground biodiversity conservation.

We often fail to recognize that in addition to specific conservation or conflict management interventions such as Snow Leopard Enterprises or livestock insurance programs, building strong and resilient relationships with the community and maintaining effective communication with them is essential in securing peoples' support for conservation.

Sustained presence in the field and participation in the way of life of local communities is critical for building these relationships. While participant observation – relying on immersion of the researcher in the community within permissible limits – is a well-recognized technique in anthropology, the value of such immersion for conservation continues to remain underappreciated.

Indeed, just like the setting up of a Protected Area is a long-term commitment, so is community-based conservation. Brief or periodic community-based interventions that are not founded on a continuing and communicative relationship with local communities carry the risk of causing a mismatch of expectations between conservationists and the local community. Misguided programs can also create problems within the community by causing societal divisions, disenchantment with conservation organizations, and can, over the longer term, cause more damage than good for the biodiversity that ones seeks to protect.

When community engagement largely relies on periodic structured meetings and workshops in lieu of immersion in the community, several elements of the PARTNERS Principles tend to get compromised. Insufficient and infrequent local presence of conservationists allows only a limited and even flawed understanding of the threats that need to be addressed, leading to misdiagnosis of conservation problems, and, therefore, continuing decline in the status of biodiversity.

Insufficient and infrequent local presence of conservationists allows only a limited and even flawed understanding of the threats that need to be addressed, leading to misdiagnosis of conservation problems.

In one of our community-based conservation sites in the Gobi Altai and in another one in the Kyrgyz Tien Shan, where we had programs running for many years, we discovered far too late that the main threat to snow leopards was no longer poaching or retribution killing – which our programs were designed to address – but rather the expansion of mining operations into snow leopard habitat. Although we were able to reprioritize our activities and catalyze effective action in both situations at a short notice, these examples serve as reminders of the risk of misdiagnosis due to inadequate field presence.

Long-term and sustained presence within the community allows the creation and delivery of long-term conservation programs, and also facilitates almost every other aspect of the PARTNERS Principles. Sustained field presence serves as an early warning system when new threats to conservation emerge, or when there are societal developments that can damage conservation efforts unless they are adapted appropriately to the changing situation (Chapter 8: RESPONSIVENESS).

Constant interaction with local people as fellow human beings improves the ability of the conservationist to understand the community's constraints and outlooks, and the hardships they face (Chapter 7: EMPATHY). It better enables the conservationist to relate to community members in an equal and respectful way (Chapter 4: RESPECT), rather than viewing them as, at best, a stakeholder in conservation, or a recipient of conservation aid. Or, at worst, as the "other side", the root cause of conservation problems.

Immersion in the community also helps provide a deeper awareness and understanding of the local socio-political and cultural situation, the social capital and the key threats to biodiversity, thereby enabling the design of contextually relevant conservation initiatives (Chapter 3: APTNESS). An understanding of local political and societal aspects, and perhaps most importantly, the development of mutual trust that sustained presence makes possible, help create integrative initiatives based on mutual interests, rather than distributive ones based on positional bargaining (Chapter 6: NEGOTIATION).

Indeed, it is not possible, or even necessary, for the practitioner – or the anthropologist – to be present in each community all the time. In my experience, the conservationist being based, long-term, in any of the communities inhabiting the focal landscape of interest, and periodically visiting other communities, appears sufficient in building strong relationship and trust. Other things being similar, it helps to be based in the larger of the communities because of the greater number of people one can reach directly.

Training and hiring individuals drawn from the local communities can really help strengthen local presence, bring immense knowledge, and add value to the team (see more discussion on hiring locally in Chapter 5: TRANSPARENCY). Over time, such individuals must be supported and empowered to be able to run community-based conservation programs on their own.

Importantly, while hiring locally is an effective way to strengthen conservation efforts and make them sustainable, it doesn't absolve the practitioner from the need for immersion. Certainly not for many years, until strong relationships and adequate local capacities have been built. Researchers, who are often based in the field for extended periods of time, can also play a vital role as agents for conservation.

> The key bottleneck for community-based efforts to be effective, that often goes unrecognized, is not so much the participation of local communities, but rather, the extent of participation of the conservation organizations themselves.

Interestingly, the discourse on community-based conservation focuses a lot on the importance of ensuring the participation of local communities at various levels of conservation planning and implementation. A few useful frameworks for community-engagement – such as Participatory Rural Appraisal and its modifications like the Appreciative Participatory Planning and Action (Jackson and Wangchuk 2001) – have been advocated. Yet, the key bottleneck for community-based efforts to be effective that often goes unrecognized is not so much the participation of local communities, but rather, the extent of participation of the conservation organizations themselves. A strong and resilient relationship with local communities is the cornerstone of effective conservation, and unless there is a long-term presence, the relationships between communities and conservationists will remain weak.

The idea and the individual

As practicing conservationists, we tend to focus on the biological, economic, and development aspects while engaging in community-based conservation. A report to a funding agency, for instance, typically lists criteria such as the number of people that benefited from a community-based intervention, the additional income the participants generated, the extent of anti-poaching activities that the community undertook, the extent of threat reduction, or biological responses such as the increased population of the target species. Project performance indicators also tend to include metrics such as the size and number of meetings held with local communities and other stakeholders.

Useful as these indicators are, they ignore the role of emotion and relationships in community based conservation. It is helpful to keep in mind that a community is made up of individuals, with emotions, perceptions, and worldviews different from each other and from those of the conservationist trying to effect change. In pushing the conservation agenda, one is, knowingly or unknowingly, appealing to peoples' emotions.

Individuals in the community will support a conservation program not just because they stand to personally and directly gain from it. In fact, in many community-based initiatives, including some of ours such as the village reserve, the gain at the individual level is diffused compared to the more tangible benefits at the level of the community. Individuals in the community will often choose to support a program – or oppose it – because of emotion, an under-recognized aspect in conservation.

At an informal gathering over tea in the village of Kibber, which was the first to start a community-based livestock insurance program in 2002, Chhering Tandup Makhan reminisced his feelings from more than a decade earlier. Makhan was one of the key local people who had helped initiate the program. He recounted with pride, amusement, and a hint of exaggeration. "I have to admit I did not understand much at all about what the insurance program would do. But Charu was a friend. I knew he meant well. So because of our friendship, I decided to support the idea without understanding it! It did eventually turn out well!"

Individuals, whether it is the conservationist or the local champion like Makhan, matter a lot in community-based efforts, not just ideas. While Makhan's recounting may point to the inadequacy in how well the idea was communicated and discussed with the community, it underscores the rarely recognized value of emotion and trust in community-based conservation. These are not easily quantified or written about in publications and technical reports. Nor do they get built over formal meetings and workshops.

Do:

- Sustained-field presence and immersion in the community
- Building strong relationships with local people
- Training and hiring local people in the conservation team

Don't:

- Forget that people's emotions can be as important as their rational motives

Chapter 3:
APTNESS

Getting an "easy win" by replicating a community-based intervention that has proven to be effective elsewhere can help in building relationships with a new community. However, replication will often not achieve biodiversity conservation, which requires a suite of contextually appropriate conservation interventions that are designed or adapted at the community level. Multi-faceted programs with multiple interventions are generally more effective as they are able to address a diversity of threats and reach out to a wider section of the community.

The aptness of community-based interventions for any site or situation should be assessed in multiple ways:

- o Are they designed to address the main threats to biodiversity in the area?
- o Are they founded on a robust scientific understanding of the problem?
- o Is there a role identified for the entire community or its representatives in the intervention portfolio?
- o Are the interventions culturally appropriate?
- o Are they in agreement with universal values?
- o Are they designed keeping in mind the local socio-economy, social capital and skill sets?

Understanding the issue: every problem is not a nail

As conservationists, we are constantly aiming to expand the impact of our programs, or, to use typical NGO language, to upscale. Replicating a successful intervention in other sites is a standard and obvious way of upscaling. Sometimes this can work well (but see later discussion on the value of multi-faceted programs).

For example, instances of depredation by snow leopards inside corrals usually result in multiple livestock kills, causing the farmer high financial losses. In an area where such damage is frequent, collaborative predator proofing of corrals can really help the farmer and is a potentially useful way to garner community support for conservation. It is also likely that the same approach, with minor adaptations, would work well in other communities where livestock depredation inside corrals is a major issue.

Indeed, such interventions that have relatively wider relevance and are easy to replicate, can play a useful role in initiating or strengthening communication and relationships with communities. However, the problem begins if we start assuming that they are adequate to address the key threats to biodiversity or focal species in every site.

It is useful to keep in mind that while the replication of a successful community-based intervention in other sites can be useful, sometimes it may be only partly useful, at times a waste of conservation resources, and at worst, damaging for the society or biodiversity. Yet, replication remains an invariably enticing way of planning an expansion for a variety of understandable reasons.

Interventions that have relatively wider relevance and are easy to replicate, can play a useful role in initiating or strengthening communication and relationships with local communities. However, the problem begins if we start assuming that they are adequate to address the key threats to biodiversity or focal species in every site.

Community-based conservation is time-consuming and challenging, and when an intervention works well after years of effort, the desire to replicate it in other sites is natural. It is also convenient to plan and propose the upscaling of ongoing programs in terms of interventions rather than measurable impacts on biodiversity, considering the complexities, time lags, and logistics involved in biological responses and their measurement.

From the perspective of funding agencies looking for tangible project impact, especially those sensitive to human issues, a focus on the intervention helps us draft more quantifiable indicators. And the fact that the intervention has already been tried and tested elsewhere instills confidence in its potential positive impacts for people and biodiversity.

It is also human nature and scientific motivation to look for general if not universal answers and solutions. Or, in the case of the conservationist, an innate desire for a conservation panacea.

The process of community-based work, like any other, also tends to concentrate the attention of the practitioner on the interventions. In implementing a conservation intervention, over time, like in any other initiative, the initial excitement of starting or expanding the program can get taken over by the routine tasks essential to keep the programs going. As this happens, the process indicators of the interventions – such as the number of people participating, the extent of benefit they are deriving, or the number of meetings conducted etc.– can take over as the guiding force at the cost of the larger vision and ultimate purpose of the effort: securing the status of biodiversity.

As the approach becomes narrowly focused on the interventions and their replication, we are often setting ourselves up to be surprised. When we actually begin to measure the impact on biodiversity, conservation-related behavior of people that the program aims to influence, or even the attitudes of the target community, we get unexpected results.

We may find that the state of biodiversity continues to degrade and the threats to conservation continue to intensify and diversify. Yet, in parallel, the process indicators – such as the number of people benefiting from the intervention economically, the number of community meetings etc. – might continue to convey that the program is running well.

Such a paradoxical situation arises not necessarily because the program is not implemented properly or because the conservation vision has taken a back seat. It may arise because the problem itself may have been misdiagnosed in the first place, or the intervention might be contextually inappropriate.

To give an example, after successfully piloting it in Mongolia, we began running the Snow Leopard Enterprises (Chapter 10: Snow Leopard Enterprises) program in Kyrgyzstan many years ago. Snow Leopard Enterprises (SLE) is designed such that in exchange for opportunities for livelihood enhancement, local communities agree not to engage in illegal hunting, and to actively prevent poaching by outsiders in their areas of grazing and resource use. It took us several years to understand that while SLE did have a positive impact with the communities living around Sarychat Reserve in the Kyrgyz Tien Shan Mountains, it was unable to reduce the extent of poaching by outsiders.

In retrospect, we realized that it was far-fetched – and even unfair – on our part to have expected the local community to prevent poaching by outsiders, considering that the latter are usually influential and politically well connected, while the local people in this case do not have ownership or rights over the land they use for grazing livestock. In fact, they have to rent the grazing land from landowners living elsewhere around Lake Issykul. Not surprisingly, while they were able to honor part of their commitment by not poaching themselves, the local communities were unable to prevent poachers from elsewhere.

The inadequacy of SLE to comprehensively address the issue of poaching in the Tien Shan is an example of how an intervention that enjoyed reasonable success elsewhere was contextually inadequate to address a rather similar problem in a different societal context. In response, we have since initiated a Citizen-Ranger Wildlife Protection Program, which runs in parallel with SLE. A collaborative effort with INTERPOL and the Kyrgyz government, this program trains and motivates rangers and local community members to work together to apprehend poachers.

Mark Twain is famously, though perhaps apocryphally, said to have written, "To a man who has a hammer, every problem looks like a nail." As practicing conservationists, we too carry the heavy burden of the metaphorical hammer. We tend to focus excessively on and celebrate our interventions, while neglecting the complexity and uniqueness of conservation problems. We tend to disregard the variation in the underlying societal dynamics in different sites or at different points in time. It would help considerably if community-based conservation thinking could shift its emphasis from 'what' to 'why'.

Addressing the problem: every solution is not a hammer

Unfortunately, there is no conservation panacea, and certainly not for community-based conservation. In fact, the complexity and dynamism of conservation threats and societal dynamics have prompted environmental issues to be labeled wicked problems, which, in a manner of speaking, have no solution (Ludwig et al. 2001). A term borrowed from social planning, a wicked problem is one that is unique; without definitive formulations, stopping rules or solutions; is constantly changing, and can be considered a symptom of another problem (Rittel and Weber 1973).

The experience with our first village wildlife reserve in Spiti Valley, India, where we worked with the local community to free up some of the land from livestock grazing to enable wild ungulate recovery, is a somber and, at the same time, somewhat amusing experience to consider.

One of our underlying assumptions had been that wild ungulate population recovery would deflect some of the carnivore predation away from livestock to wild prey (Mishra et al. 2003a). A decade later, our own research invalidated the assumption. It showed that wild ungulate abundance was the main determinant of snow leopard abundance, and that an increase in wild prey could actually cause an increase in the extent of predation on livestock, rather than a decrease (Suryawanshi et al. 2013).

The village reserve effort did result in a four-fold increase in wild ungulate abundance. So from the perspective of conservation, our collaborative effort had succeeded in enabling wild ungulate recovery, and perhaps even facilitating the use of the area by snow leopards (Mishra et al. 2016a), both rather desirable outcomes.

However, from the community perspective, it is unlikely that the village reserve helped in reducing livestock depredation. On the contrary, it led to a new issue, that of more crop depredation by the wild ungulates. To make matters worse, this occurred during a period of rapid socio-economic transition that saw crops largely replace livestock in their relative importance in the local economy (Mishra 2000).

Fortunately, we had also started an insurance program with the community in question to address the problem of livestock depredation, which has been running well. More recently, we have had to initiate discussions and pilot new interventions to address the issue of crop depredation.

It is therefore instructive to consider the inherently wicked nature of conservation problems. Doing so helps us realize that seemingly identical conservation problems can actually be very different, and, even when they are similar, the conservation interventions required may vary from one site to the other. And perhaps most importantly, that there are no final solutions.

Assessing aptness

The recognition that there is no single correct solution in community-based conservation is humbling. But it need not be a cause for despondency. On the contrary, acknowledging that no solution is perfect makes it much easier to try out new interventions. It also makes it easier to critically evaluate ongoing interventions, accept the shortcomings and adaptively evolve the programs.

The knowledge that there is no single or correct way to address a problem in community-based conservation also helps reduce the fear of making mistakes. Way too often, conservationists hesitate to try out new possibilities because of the fear of going wrong, even while fully recognizing the need for thinking and acting outside the box.

The situation, the conservation threats, the constraints, capacities, and opportunities vary between areas and between communities in the same area, and they change over time. It is therefore important for any community-based effort to be sensitive to this dynamism, and for the interventions to be designed or at least adapted to the specific contexts and communities. While no effort will be perfect or correct given the nature of the problem, the appropriateness of an intervention or a set of interventions for a situation or a community can and should be assessed in multiple ways.

The threats

Strangely, the best way of judging the aptness of specific conservation interventions - the raison d'être of conservation - tends to be amongst the most neglected. Are the interventions designed to address the key threats to the biodiversity that one is trying to preserve?

A new intervention is often designed in response to particular threats to biodiversity, and can potentially work well if informed by adequate science, supported by the community, and implemented well. However, it is useful to keep in mind that any intervention usually addresses a limited number of threats, or a limited number of aspects of any threat. In reality, biodiversity in any site tends to face a multitude of threats, and as a rule of thumb, a suite of interventions with any community tends to be more effective and resilient than a single intervention (See later discussion on multi-faceted approaches).

Furthermore, when we try to replicate the interventions in other areas, it is often based on the assumption that the threats are similar. We neglect to conduct comprehensive threats assessments, even though several simple and useful frameworks for threats assessment are available (e.g. Salafsky and Margolius 1999).

Indeed, many of the threats tend to be common or similar between sites, so, with good fortune, the same interventions can have a positive impact on biodiversity in the new sites. But they need not, especially if there are other, more overwhelming threats to biodiversity that the interventions were not designed to address.

As mentioned earlier, the potential expansion of mining operations into snow leopard habitats in our community-based sites took us by surprise on more than one occasion. This happened because our field presence was not adequate, and because we were focusing excessively on our conservation interventions instead of letting ourselves be guided by the actual threats.

Maintaining sight on the threats to biodiversity in any area is critical. These are constantly changing, and, more often than not, intensifying. Routinely evaluating whether or not our interventions are effectively addressing the main threats to biodiversity is one of the most essential and fundamental ways of judging their aptness. If they aren't, we need to adaptively improve them. If that is not enough, we need to design new interventions. In conservation, one size does not fit all.

The science

Community based conservation is more craft than science, where social sensitivity and skills of the conservationist matter as much as or more than scientific frameworks or sociological methods. Yet, it is difficult to overemphasize the importance of robust scientific understanding of sociological and ecological issues in designing, implementing, monitoring and adapting community-based efforts.

> Community based conservation is more craft than science, where social sensitivity and skills of the conservationist matter more than scientific frameworks or sociological methods. Yet, it is difficult to overemphasize the importance of robust scientific understanding of sociological and ecological issues in designing, implementing, monitoring and adapting community-based efforts.

One often experiences situations where the community is facing an issue, they have a clear idea of what is to be done, and they request for the conservationists' support. While community knowledge is to be valued highly and their solutions given high consideration, I have always found it useful to insist on first studying the problem, collaboratively if possible. Indeed, it is important to explain respectfully why developing a better understanding of the problem is required. This can be done by providing examples and competing explanations, explaining nuances, and discussing other possible options.

For instance, when people requested our support for large-scale fencing to protect crops from wild ungulates, we helped them understand how such fencing could be damaging for wildlife movement. We also helped them see that there were other options such as temporary solar fencing that could be explored, but only after mapping the hotspots for crop damage. While initiating the study to look into a long-term solution, we could simultaneously assist the community immediately by creating support for temporary guards from the community. These guards are tasked with maintaining vigil for a few months each year, when crops are most vulnerable.

More fundamentally, science has a role in defining the conservation problem that one is trying to tackle in the first place. This might sound somewhat exaggerated, but it isn't, and is better explained with an example.

In the late 1990s, when I started working in the Buddhist Trans-Himalayan region, the prevailing wisdom at the time conveyed a somewhat rosy picture of the state of wildlife. Anthropologists had written about the Changtang region of the Tibetan Plateau "...The balance of livestock, people, and pasture is not degrading or overgrazing the pastureland...There are an abundance and diversity of wild ungulates such as antelope, wild asses, gazelles, and blue sheep." (Goldstein and Beall 1989, p. 179).

Ecologists too betrayed visions of harmonious coexistence between Trans-Himalayan people and wildlife, though in a more guarded, indirect manner. "...A generally benign association (of wildlife) with a sparsely distributed population whose traditional land use and religious practices have permitted long-term coexistence" (Fox et al. 1994). Or, "Wild animals occur in low densities and need larger areas to maintain their viable populations..." (Chundawat and Rawat 1994, p.3).

It would have appeared that little, if anything at all, was needed in terms of active conservation effort. The remote, high altitude mountain landscapes, a sparse density of humans, and the prevalence of Buddhism, all conjured up and rendered plausible the notion of a Trans-Himalayan Eden.

Research, however, soon belied this pastoral idyll. The region was in the middle of a rapid socio-economic transition, and while being geographically remote, the local economy was already getting integrated with regional markets (Mishra 2000). Studies documented the serious extent of economic loss suffered by local communities due to livestock predation by snow leopards and wolves, and the retaliatory carnivore killing (Mishra 1997). Research showed that the rangelands were overstocked with livestock (Mishra et al. 2001) and populations of wild ungulates were depleted because they were outcompeted for resources by livestock (Bagchi et al. 2004, Mishra et al. 2004).

These research findings came as a surprise to many, including us. They also catalyzed us to start our community-based conservation work in the region, and led to initiatives like the village reserve and the community-based livestock insurance program. Research thus helped identify and define the conservation issues that had remained ignored.

The case of village reserves is useful to consider here once again. When we began our work, the density of blue sheep *Pseudois nayaur*, the main wild ungulate in our study area, was relatively low. Although the hunting of wild ungulates wasn't chronic, sporadic poaching instances were prevalent, a few by local people and others by defense personnel, assisted by locals (Mishra et al. 2003a).

Wild ungulate density is a key determinant of snow leopard abundance. Without the benefit of research findings, we could have easily been tempted to consider an intervention like SLE to facilitate an increase in the wild ungulate population. There is no doubt that SLE could have been useful for the local community by bringing the women an additional livelihood source, and would have also helped curtail the instances of wild ungulate hunting.

However, it was highly unlikely that SLE would have helped in facilitating any significant increase in the wild ungulate population. Research was already showing that the wild ungulate population was largely limited by excessive livestock grazing (Mishra et al. 2001, 2004); a threat that an intervention like the village reserve was able to specifically address. And one that SLE in its standard form would not have been able to do anything about.

Thus, science has a fundamental role in helping us recognize and analyze the key conservation issues and their societal underpinnings. It helps develop robust situation analyses, better identify the threats to biodiversity, and frame conservation problems appropriately.

Science can help frame measurable conservation targets, and also informs us about the kind of interventions that are more likely to help achieve them. It gives a sense of the kind of resultant responses we can expect, and time frames over which we can hope to see measurable change.

By helping identify the appropriate indicators to be measured, and by designing statistically and empirically robust frameworks for measurement and data analyses, science also plays a critical role in monitoring of program performance, a subject discussed later (see Chapter 8: RESPONSIVESS).

For now, it is useful to keep in mind that an intervention that is designed without a robust scientific understanding of the socio-ecological context is less likely to be apt. And that an expansion which doesn't begin by first developing a scientifically robust situation analysis is less likely to succeed.

The scale

The household or the community?

Operationally, individuals or individual households tend to form the actual unit of participation in many community-based conservation interventions. For instance, livestock are owned by individual households, and therefore, the participation in a livestock insurance program is at the household level. Similarly, individuals choose whether or not to participate in SLE. The number or proportion of participants or families is, therefore, considered an important metric of the reach and impact of most community-based conservation interventions.

Where individuals or households are the main units of participation, it is important that elements be specifically designed in the interventions to facilitate the potential involvement and ownership of the entire community, not just the participants. The need for this is obvious. Unless the entire community's support for conservation is generated, the interventions will not have the desired impacts on the status of biodiversity. In SLE, for instance, ensuring that a part of the funds generated are directed for community welfare or to a micro-credit program that is open to non-participants has been very helpful.

Another way to reach the entire community is for the interventions to be multi-faceted, which, in any case, tends is more desirable than single interventions for reasons discussed later (see discussion on multi-faceted approach).

It is helpful to ensure a role for the entire community or its representatives in any conservation program, even if only a proportion of the individuals are actually involved in the intervention. It is also important, as discussed earlier, that the designing or tailoring of interventions be undertaken at the scale of each community for the programs to be contextually appropriate.

When a conservation intervention is designed, or at least adapted at the scale of each community through discussions and negotiations, it is likely to experience relatively high local ownership. This ownership, as we shall see later (see Chapter 6: NEGOTIATION), is one of the important determinants of the resilience of any community-based program, and of the ease with which the interventions can be adaptively improved over time.

Yet, for all the talk of the need to involve local communities from planning to implementation of conservation programs, when it comes to actual practice, we often try and replicate the same standard prescription and set of rules we have worked out for any intervention. While we tend to do this for multiple, understandable reasons, as discussed earlier, shifting the focus from the standard intervention to the specifics of each community would help in making the programs stronger.

Small groups, or large communities?

With conservation interventions, we want to reach out to as large a proportion of the local community as is possible. The greater the number of people involved, the bigger the potential positive impact on the status of biodiversity. Most of the community-based interventions, in any case, need a minimum threshold number of participants to be effective and sustainable.

While larger groups are desirable, sometimes this can become problematic. Many interventions such as the livestock-insurance program or village reserves rely on the willingness of the participants to cooperate with each other. As the group size increases, the willingness to cooperate can decline. This is not just because of the reduced communication among participants as their number increases.

Game theory suggests that the rate of cooperation in an interaction is inversely related to the number of people involved, and the potential rewards from cheating tend to increase with group size (Colman 1999). Note that the term cheating is used here purely in a decision-theoretic sense, denoting logical action explained by rational self-interest, and has no moral connotation.

In practice, this can manifest itself as, say, an increasing tendency for filing false claims in a livestock insurance program. A few years ago, there was an instance of a false insurance claim detected in one of our programs in Spiti Valley, Western Trans-Himalaya. Because the community was relatively small with around 50 participant families who all knew each other, the transgression was easily detected, and the claimant let off with a warning.

The ideal size of the group for a particular community-based intervention may vary with the community and the intervention, and it is difficult to specify numbers with which to work. Instead, this is best judged by the conservationist together with the community members.

When communities and potential number of participants are large, it may be useful to manage the interventions at the scale of traditional administrative groupings rather than of the entire community.

Communities are well aware of these issues. Beyond a certain size, most local communities traditionally divide themselves into smaller groups for ease of internal administration and management. In snow leopard landscapes, community groups are usually based on how close the houses or gers are to each other, or the proximity of their resource use areas. There are interesting systems of decision-making within each group, often democratic, with group leaders being responsible for negotiation and coordination between groups and eventual community-level decisions (e.g. Mishra et al. 2003b).

When communities and potential number of participants are large, it may be useful to manage the interventions at the scale of traditional administrative groupings rather than of the entire community. It is preferable to rely on traditional administrative groups – provided they are voluntary and equitable – rather than create new ones, since any imposed grouping may be less efficient and could also have unanticipated consequences for community-cohesion.

There is another hidden lesson here. Working with smaller, tightly functioning community units has advantages as discussed. Further, beyond the management unit, sometimes one just needs to take one small step at a time, especially when trying out new initiatives (see Chapter 5: TRANSPARENCY). Conservationists sometimes hesitate to try out new ideas because they feel their effort is too small, impacting a limited area, or involving only a few participants. In community-based conservation, no step is too small.

Small areas, or large landscapes?

While small has advantages, it can also be a rather important and prevalent problem when it comes to the spatial coverage of community-based conservation efforts.

Our ability to work with any given community tends to be influenced by multiple factors such as our familiarity, presence, relationship, ease of access, wildlife value of community-land, specific threats to wildlife, resources and manpower. The inclusion of communities in conservation efforts within a landscape, therefore, is rarely uniform. This is especially the case when community-based efforts are undertaken without a geographical reference such as a Protected Area or some unit of conservation in mind.

Community-based efforts implicitly focus on and tend to be constrained by the size of the area or habitat owned or influenced by the community in question. On the other hand, common sense – as well as the island biogeography theory – underscores the importance of protecting larger habitat patches rather than smaller ones for effective conservation and to reduce the chances of local extinction of species. There is often a mismatch between the area of influence of the community and the habitat needs of the species or biological assemblage that one is trying to conserve.

For instance, the home range size of an individual snow leopard, our conservation flagship, can be spread over a few hundred square kilometers. The area of influence of communities in snow leopard habitats, on the other hand, typically varies from a few tens to a few hundred square kilometers, and, rarely, a few thousand.

On average, therefore, in a given community, the habitat area that could potentially be protected through community-based efforts will often be smaller than the home range size of even a single snow leopard. This means that even if community-based efforts managed to result in strict protection in one community's area, a snow leopard that uses this habitat, in its normal course of movement, could still get killed in another community's area.

When wildlife includes landscape species like the snow leopard, the spatial area of reference cannot be the community land, or even a protected area alone. The focus has to be on entire landscapes that can support breeding populations, or form important biological corridors connecting other populations.

Therefore, when wildlife includes landscape species like the snow leopard, the spatial area of reference cannot be the community land, or even a protected area alone. The focus has to be on entire landscapes that can support breeding populations, or form important biological corridors connecting other populations.

In such a landscape-level approach, while the focus is on promoting conservation across entire landscapes, the community or a cluster of neighboring communities still remains the operational unit of conservation. The spatial units over which conservation actions are implemented (and some of the biological responses are measured) are usually delineated based on a conflation of ecological, geographic, threats-related, and administrative factors.

These and other issues such as the important role of the Government and multi-sectorial cooperation for landscape scale community-based conservation are discussed later (Chapter 9: STRATEGIC SUPPORT). For now, it is useful to keep in mind that our objectives and the biology of the species involved play a key role in deciding the scale-aptness of community-based conservation programs.

Socio-cultural aptness and value orientation

Value orientation

Assessing the cultural appropriateness of any community-based intervention is essential. Culture represents a complex of beliefs, practices, norms, values, and symbols (Schwartz 2006). An important aspect of cultures is defined by the value orientations of people – their shared ideas of what is good or desirable – which must be considered in community-based conservation.

For instance, to many researchers and conservationists trained in western style wildlife management, trophy hunting of wildlife is a perfectly legitimate conservation tool, provided it is implemented well. Indeed, while mismanaged trophy hunting has contributed to depletion of wildlife (the Kyrgyz Republic, consequently, had put a temporary moratorium on trophy hunting across the country), in other areas, such as northern Pakistan, well-managed community-based trophy hunting has arguably helped both people and wildlife in snow leopard habitats.

A Buddhist monk, or even the average practicing Buddhist herder living in snow leopard habitats, however, is likely to find the idea of taking life and inflicting insufferable pain on another living being – for sport – preposterous and deeply sinful. Ironically, their antithetical views notwithstanding, the monk and the trophy hunter may fully share a common concern for protecting other forms of life.

It is these value orientations that make an intervention like trophy hunting potentially apt in say Islamic or animistic communities that have retained a strong tradition of hunting, while rendering it totally inappropriate in others. Value orientations of the local community are essential to consider, but the conservationist must also be mindful of those of the larger society supporting conservation. Consequentialist reasoning – whereby the ends are seen to justify the means – can be problematic, particularly when it comes to conservation tools such as trophy hunting (Nelson et al. 2016). Being aware of these issues is essential. Such awareness comes with deeper thought and sensitivity.

Values

The need to consider value orientations is an important issue, but not the only one. The sensitivity toward contextual value orientations may need to be balanced with the need to uphold certain universal values. For instance, too often, social realities dictate that in community-based conservation, inadvertently, we end up largely working with the male members of the community, who are seen to be making the decisions at both the household and the community levels.

This can be problematic not just from the perspective of gender equity. For instance, recent research indicates that women tend to have greater negative attitudes towards predators such as the snow leopard, partly because they might bear disproportionately greater costs of wildlife-caused damage such as livestock depredation (Suryawanshi et al. 2014). Unless community-based programs have an adequate focus on women and ensure their representation, building societal support for wildlife conservation will remain difficult.

In response to this felt need, in Pakistan, our Ecosystem Health Program expanded its representation to include and train women as community livestock extension workers. In India, SLE was added to the repertoire of community-based interventions to specifically seek the support and involvement of women in snow leopard conservation.

The need for gender representation, however, applies not just to the communities. Working effectively with local communities also requires that the team of conservationists have adequate gender representation.

The developments in our vaccination program in Pakistan, or expansion of SLE in India, would have been difficult without the presence of highly capable women in our staff. We have been very encouraged and excited by the enthusiastic response of the communities in India, where our women staff have led the piloting of the

SLE initiative. This enthusiasm of local women is in stark contrast to their rather indifferent response more than a decade back when I first discussed SLE with them. There were no women conservationists in our team then.

Thus, community-based efforts need to strike a balance between the contextual value orientations of specific cultures and certain universal values, such as gender and social equity. These judgments are perhaps best made based on intuition and common sense.

Socio-economics and social capital

Local economies and skill-sets

The aptness and performance of any conservation intervention in a given community will often depend on the local socio-economy. It is therefore useful to assess the socio-economic status of households at the community level while designing or adapting a community-based intervention.

For instance, the ability of livestock owners to contribute premium into the insurance fund depends on their economic status. In relatively poor communities, this ability will remain limited. Therefore, a greater proportion of the insurance fund may need to be subsidized through conservation funding compared to relatively wealthy communities, where participants may have the willingness and ability to pay relatively higher premium amounts.

Similarly, in otherwise comparable circumstances, the willingness to participate in an income generation program like SLE, and its potential impact could be expected to be higher among relatively less affluent communities, as any additional income for them would form a much larger proportion of the average household income. In more affluent communities, the addition to the income may need to be much higher for the program to have the desired level of participation and impact.

It is also useful to keep in mind the variation in skill sets and social capital amongst communities, which can have implications for the performance of community-based interventions. While interventions are best designed keeping local resources and available skills in the community in mind, almost invariably, the skill sets need periodic enhancement. Given the variation in existing skills, the need for training again can vary between communities. For instance, people in relatively remote communities may have high natural history knowledge, while a community-based tourism intervention might require greater training emphasis on housekeeping and hospitality. In contrast, training for people living in more integrated communities or setups like townships may require a greater emphasis on natural history and nature interpretation skills compared to hospitality.

> While interventions are best designed keeping local resources and available skills in the community in mind, almost invariably, the skill sets need periodic enhancement.

Social capital

The concept of social capital recognizes the value of social networks, trust and norms as a resource for action and for addressing individual or collective problems (Coleman 1986). Social capital can be a diffuse but critical element in the effective delivery of any community based program. It is especially useful in the implementation of collective agreements and norms, so that community-based interventions can actually lead to an improvement in the status of biodiversity.

Communities may differ in their available social capital, a useful measure of community coherence. This can have implications not just for the ultimate performance of community-based programs, but also on the aptness of any community-based intervention.

Some interventions depend more on social capital compared to others. For instance, our SLE and livestock vaccination programs (Chapters 10 and 12, respectively) rely predominantly on individual involvement, and less so on social capital, although the latter continues to have an important role in ensuring conservation compliance. In comparison, the livestock insurance program (Chapter 11) has a much greater reliance on social capital. It requires systems and norms that can ensure that the participants pay their premiums on time, and that the committee members manage the funds with integrity. It requires social networks and trust to deal with moral hazards and to deter the temptation to file false claims. And it depends on voluntary contribution of time and effort of insurance committee members to run the program.

Thus, the available and required social capital can help assess the suitability of any community-based intervention. It is also useful to keep in mind that social capital needs resource investment, both economic and cultural (Portes 1998). Sometimes, in the interest of long-term and sustainable conservation outcomes, it becomes useful to invest in enhancing it, a subject discussed elsewhere (Chapter 8: RESPONSIVENESS).

Interestingly, while community-based conservation interventions depend on it, they can also contribute to enhancing the available social capital. The Youth Council of Kibber village is a good example. This was a loose collection of young villagers brought together by their shared love for having a good time, mostly a good game of cricket! Yet, this group became more organized, and indeed more responsible, when they started playing a role in community-based conservation interventions.

They acquired a fair amount of respect within the community especially after effectively managing the livestock insurance program in the village – to the extent that, when internal conflicts led to a temporary breakdown of the traditional village administration system, this group, by now respected and seen as non-factional, was requested to administer the affairs of the village. They ended up doing so for the nearly two years it took for the disputes to be resolved and status quo to be restored.

Social capital, and the aptness of specific conservation interventions for a given community, therefore, can change over time. Often, timing is critical in community-based conservation, as we shall see later (see Chapter 6: NEGOTIATION and Chapter 8: RESPONSIVENESS).

Multi-faceted approach

Diversity is ingrained in our thinking. The stability of ecosystems is thought to increase with the diversity of its component species (McCann, 2000). Peace and stability among nations is suggested to increase with the diversity and extent of their economic relations and interdependence (Gartzke et al. 2001). Similarly, the effectiveness of community-based conservation efforts, and the resilience of the relationship between communities and conservationists, is influenced by the diversity of interventions – for several reasons.

As discussed earlier, a single community-based intervention is rarely able to address all the key threats to biodiversity in an area (see section above on Threats). This simple recognition gives rise to the rule of thumb that multi-pronged approaches are better than single interventions.

In fact, even a single threat often has multiple dimensions. To take an example, let us consider the seemingly straightforward threat of retaliatory killing of predators in response to predation on livestock. Human attitudes and behavior underlying different responses to predators, such as retaliatory killing, can be influenced by range of factors. These include individual human experiences and attributes, socio-economic indices of the family and the community, relationship with the state, the appearance and behavior of the carnivore itself, and so on (Suryawanshi et al. 2014).

This creates the need for management efforts to be made multi-pronged. For example, a conservation approach needs to be able to address at least three aspects to be effective in the case of livestock predation and retaliatory killing (Mishra and Suryawanshi 2015, Mishra et al. 2016b).

- Steps to reduce livestock losses through better livestock protection,
- Mechanisms to share and offset economic losses when livestock depredation does take place,
- Interventions to improve the social carrying capacity for the predators through livelihood enhancement and awareness programs.

All three aspects require different kinds of interventions. For instance, predator-proofing of corrals in certain sites, and building incentives for better herding in others (in situations where losses occur while livestock is grazing in the pastures), can help with better livestock protection. But neither of these interventions is designed to help offset depredation costs, for which an intervention like the insurance program needs to be considered (Chapter 11). Insurance programs themselves can create a moral hazard (see discussion above on Social Capital), and therefore, thinking of ways to reward better herding can be useful. Similarly, Snow Leopard Enterprises (Chapter 10) or collaborative veterinary care (Chapter 12), where appropriate, can help with livelihood enhancement, but not with better livestock protection.

There are other reasons that make a multi-faceted approach more meaningful. For community-based efforts to lead to effective biodiversity conservation, it is important to have the willingness and support of the entire community. This underscores the importance of involving as many households as possible in the program. Single interventions rarely manage to reach out to the entire community.

For example, there are households that do not have livestock, and would not benefit from interventions like corral improvement or livestock insurance programs. Yet, it is important to involve them, because their actions could be equally or more conservation-unfriendly compared to the participants. On the other hand, these households could become potentially useful conservation allies if they are involved in a meaningful way.

> Often, the needs of the community rather than those of biodiversity, or their conceptual familiarity with an intervention, influence their readiness to pilot it.

Diversification of interventions, therefore, helps in making the programs more inclusive. Families without livestock, for example, could perhaps be involved through other contextually meaningful programs such as SLE. The potential inequity in the community created by incomplete coverage of conservation programs could, in fact, lead to problems for both the society and the biodiversity.

It is important to keep in mind, however, that it is neither prudent nor logistically feasible to start multiple interventions in a community at the same time. Diversification must take place step by step.

Starting with interventions that address the main threats to biodiversity and diversifying over time is useful whenever possible, but not always feasible. Often, the needs of the community, rather than those of biodiversity (see Chapter 8: RESPONSIVENESS), or their conceptual familiarity with an intervention, influence their readiness to pilot it. This is okay as long as it is regarded as an initial step in the larger conservation vision for the area, and the program is diversified over time to address the main threats to biodiversity. When a community is able to run an intervention appropriately for some time, it tends to become relatively more open as well as more capable to experiment with others.

A multi-faceted program developed over time improves the resilience of conservation partnerships with local communities. A relationship based on a single intervention will collapse if the intervention were to fail for some reason – or in some cases even if it were to succeed. For instance, there is little to do in terms of follow-up in an intervention such as collaborative predator-proofing of corrals. Once the corrals are improved, there is no tangible avenue left for sustained engagement with the community, and for encouraging conservation-friendly behavior.

Indeed, when local communities are convinced of the long-term interest, presence, and potential for a diversity of engagements with conservationists, they also tend to be more responsive, and more open to adaptive improvement of interventions. A diversity of potential interventions also helps shift the interaction from positional bargaining to a more collaborative partnership (see Chapter 6: NEGOTIATION). In community-based interventions, in diversity, there is strength.

Dos:

- Assessing threats to biodiversity rather than assuming them
- Designing interventions to address specific and relevant threats
- Designing interventions that are contextually appropriate for the target community
- Working with women and ensuring adequate representation in the conservation team
- Reaching out to majority of the community, but working with relatively smaller groups
- Investing in enhancement of social capital

Don'ts:

- Ignoring social and cultural contexts when implementing programs
- Focusing solely on program participants and forgetting to build in a role for the entire community in the intervention portfolio
- Creating new groups within the community for program operations, instead of using traditional ones
- Focusing solely on individual community land for landscape species conservation

Chapter 4:
RESPECT

Respecting people's dignity is a basic human imperative. Treating local communities ethically is particularly critical in community-based conservation, where local people can become instruments to an end. A respectful and ethical stance goes beyond our conduct and civility in interactions with them. It begins with one's internal psychological orientation.

An equal partnership implies that conservationists do not view local communities as recipients of aid, and themselves as the providers. Community-based conservation efforts must follow the code of beneficence and non-malfeasance, and individual and societal differences or divisions should never be used for pushing the conservation agenda.

Basic dignity

Local communities comprise people like us, who have fundamental dignity, like we do. This is particularly critical when it comes to the participants of community-based efforts, since engaging with them is more than a value-neutral social interaction.

Local people become an instrument to achieve what is primarily, or at least proximately, our goal – the conservation of biodiversity. Although the desire, and often the outcome, is that they benefit from community-based conservation interventions, the conflicting impulse of viewing people as instruments to an end on one hand, and respecting general ethical norms of interacting with people on the other hand, can create an object-subject tension (Cohen 2001). Such issues make the ethics of community-based conservation rather important to consider.

It is useful to share an example. This involved an instance where a snow leopard was killed a few years ago, inside a livestock corral, in one of our focal landscapes in the Trans-Himalayas.

Long after the forest department had investigated the incident, a few members of our team visited the hamlet. They were tasked to determine the circumstances under which the killing took place, and, based on what they'd find, conduct discussions with the community members and report to us their collective thoughts on what could be done to prevent such incidences from recurring. Our team members met the herder who had lost many livestock in the incident, and, with the help of others, had killed the snow leopard.

Our team, on this occasion, happened to comprise young and relatively inexperienced staff. Additionally, I suspect we failed to brief them sufficiently before they left. While talking to the herder, they decided to suppress information about themselves, and to not disclose their actual purpose. They feigned marginal interest in the snow leopard incident. They felt that doing so would help them get more accurate information, and that disclosing their actual identity would deter the herder from sharing details.

Our team did manage to gather detailed and accurate information regarding the circumstances under which the snow leopard was killed. And they could do so over cups of tea, enjoying the hospitality of the unsuspecting herder, inside the comfort of his house.

There was no malice involved in what our staff did. The intention was never to deceive the herder, nor was he harmed in any way due to our actions. We could eventually start a conservation partnership with the community. We entered into an agreement, and collaboratively predator-proofed all the corrals in the hamlet. We now have a strong and continued relationship with the residents of this hamlet.

From an ethical perspective, however, our initial interactions and actions were rather questionable. They represented a classic manifestation of the object-subject tension to which I will return shortly.

It is critical in community-based conservation that we interact with local communities with fairness and honesty (also see chapter 4: TRANSPARENCY). It requires us to have a respectful stance. It requires that we refrain from deception or coercion. It requires viewing local people as equal partners, which also implies respecting their autonomy.

A respectful stance entails aspects of external behavior, but also our deeper attitudes. It is not just about external conduct and civility in interactions, but one's internal psychological orientation towards the other party (Cohen, 2001). This orientation, knowingly or unknowingly, can have a considerable influence on behavior.

Looking back, we were unfair and even deceitful – albeit inadvertently – to the herder. We did what we believed at that point was – correctly or not – best suited to meet our objective, and compromised on standard ethical norms of interacting with others. The knowledge that the herder had killed a snow leopard had presumably influenced our internal orientations, our stance, biasing our actions.

In Cohen's (2001, p. 750) words, the challenge lies in "...seeing the fundamental dignity of people despite their instrumentality." The challenge for the conservationist in community-based conservation can be even more testing. Seeing the fundamental dignity of a person despite their occasional, seemingly antithetical – and even illegal – behavior, such as killing a snow leopard, can be difficult.

Provider or recipient?

"We did so much for them, but they still did not stop hunting." One occasionally hears this sentence, or its variants, from a wildlife manager or a conservationist who has tried community-based work. Words expressing frustration. Words that communicate a feeling of betrayal.

The problem here is not only that a community-based program did not have the desired effect on biodiversity, perhaps because it wasn't designed or implemented well. Or that it may not have worked due to some extraneous factors, despite being planned and implemented well. When programs don't have the desired effect, frustration is understandable. But there is more to it.

There is an underlying stance problem here. When programs don't have the desired effect, or even when they do, the nature of our work is such that it is easy to start viewing ourselves as the provider, and the community as the recipient.

Such a stance probably arises because conservationists and managers help bring resources into an area or a community. Resource mobilization and expenditure, in fact, form a significant part of doing conservation. Considerable societal funding is spent in what could be construed as community aid.

We spend time and effort finding those resources. We work hard to use them diligently, and to implement conservation programs. We invest effort accounting for resources, and reporting back expenditure and progress – or lack thereof – to our governments or funding agencies. This reality of the importance of funding, and the role of the conservationist as its conduit, can lead to an implicit donor-recipient hierarchy. It compromises the sense of equality that should characterize our relationship with the local communities, and can even render us paternalistic towards them.

It is helpful, and even humbling, to consider that in many ways, the communities are the main provider in this interaction, in the form of their potential support for biodiversity conservation that we are seeking.

If our stance makes us view local communities as the recipient in community-based conservation, there will be no equality in the conservation partnership. There will be no fairness. This is a problem, as the very starting point of community-based conservation, alongside pragmatism, is the pursuit of fairness.

It is helpful, and even humbling, to consider that in many ways, the communities are the main provider in this interaction, in the form of their potential support for biodiversity conservation that we are seeking. We try to achieve this by supporting and empowering the local communities. Community-based conservation relies on the devolution of conservation responsibility to local people.

Thus, in most ways, in community-based conservation, we are the recipients – or, at best, catalysts for change – who depend on the community to meet conservation goals.

Our ability to view the community as an equal partner in conservation is critical in community-based work. Not just in the civility of our interactions with them, but, importantly, in our fundamental attitude. If our psychological orientation tends to view the community as the recipient, we would have lost the plot of community-based conservation even before we began.

Respecting discord and avoiding harm

Many cohesive forces keep the members of a local community together. There is the shared space, resources, history, kinship, interdependence and reciprocity, local institutions, and traditions and rituals that keep people together. This social capital forms a most important resource for effective community-based conservation.

Yet, local communities are not homogenous bodies. Like in any community, there are social divisions, class divisions, political divisions, and individual disputes. Sometimes, such internal divisions impede the progress of community-based work, even though conservationists may not be responsible for them or involved in them in any way. At other times, we may be drawn into these disagreements.

It is no doubt useful, and even important, to be aware of local divisions and disputes, because, though unrelated, they can have unintended consequences in retarding the progress of our work. Being aware of them helps to better decide on the social institutions to partner with, and the individuals to talk to (see Chapter 6: NEGOTIATION).

When faced with opposition from a section of the community, however, we are sometimes tempted to make use of the local divisions, especially political ones, to address dissent. This temptation is to be avoided at all costs. It is important that working with communities, we respect the value of human relationships, and try to ensure that our work does not weaken existing relationships among people. It is also useful to keep in mind that communities have their own systems and arbitration mechanisms for resolving disputes.

Attempts to use differences within the community for conservation would be unethical, and constitute an undue intrusion. Indeed, beneficence and non-malfeasance form important twin guidelines of any community-based work (Gambrill 2012). We must try to ensure that to the extent possible, our community-based conservation actions are able to help local communities and do not knowingly cause them harm.

Patience, communication and negotiation are the only way to circumvent such road bumps in community-based work created by internal disputes (see Chapter 6: NEGOTIATION). Getting into divisive haggling or into alliances based on local power equations must be avoided. Apart from being unethical, this strategy yields only short-term gains at best – but those tend to come at a large long-term cost to conservation efforts. Power can change sides quickly.

There are also instances, especially when there is a long-term relationship with the community, when conservationists may be requested by the community for advice, or even mediation, in settling internal disagreements. Under such situations, it becomes important to share opinions in a neutral manner, while reminding the community of our core competency, which is not in arbitration.

Factionalism, discrimination and favoritism are damaging for community-based conservation. On the other hand, maintaining neutrality and equitability, following fair and transparent processes, and if possible, promoting social justice become our responsibilities.

Dos:

- Treating community members with respect
- Seeking to create an equal partnership with the community
- Engaging in open and honest communication
- Taking note of societal divisions and individual differences within the community

Don'ts:

- Viewing local communities as recipients of aid, rather than as providers of conservation services
- Using societal divisions and individual differences within the community to advance the conservation agenda

Chapter 5: TRANSPARENCY

In an equal conservation partnership, there is no room for deceit or withholding information. It is the conservationist's responsibility to clearly communicate conservation goals to community members, explain why certain choices are made and what effects they might have. It is important that community members are involved in making choices, from conservation interventions to hiring of local staff. Community members must be provided with opportunities and avenues to seek explanations and share their advice and misgivings regarding conservation programs. Transparency requires that communication be maintained not just with community leaders or local program coordinators, but the average community member.

Transparency has various dimensions, and has high importance in community-based conservation efforts. It is ethically desirable, helps in improving program efficiency and adaptive improvement of interventions, building trust, and avoiding factionalism and favoritism.

The ethical perspective

In the most general sense, transparency implies disclosure of our purpose, and clear communication of our goals to the community. It is important that we initiate discussions with community members by declaring – and periodically reiterating – our main purpose, which is to promote biodiversity conservation with their support and involvement. Not quite the way we initiated our discussions with the herder who had killed a snow leopard (see Chapter 4: RESPECT).

Transparency implies there is no room for deceit. It also renders unacceptable the deliberate withholding of information regarding the interventions, especially their potential weaknesses and uncertainties.

It is also useful, where appropriate, to openly discuss potential conditions of discontinuation or failure of interventions, partly because such a situation can otherwise lead to discontent, and even cause internal divisions. Transparency implies that any potential negative impacts of the intervention on the community be clearly stated and discussed, in the real spirit of non-malfeasance.

As discussed earlier, communities are not homogeneous entities. There are power imbalances among people, and there is always the reasonable possibility that community-based conservation efforts benefit some people more than others. Transparency in the process and interventions can help achieve greater participation and equitability in the distribution of responsibilities and benefits among community members (also see Chapter 6: NEGOTIATION).

This implies that periodic communication be maintained not just with community leaders or local program coordinators, but also the average program participant, and even the non-participant community member. These interactions need not always be formal, and, in fact, tend to be more productive in informal settings.

Another important aspect of transparency is that community members should be able to make suggestions and share their misgivings, either in a group or even individually. They should be able to do so without any fear of reprisal. If warranted, their confidentiality needs to be respected – paradoxically in the interest of transparency.

Community members should be able to seek answers and explanations from us regarding the conservation programs. We have a professional obligation to share information with them, and being communicative and approachable is a fundamental step in enabling effective information exchange and accountability.

Transparency in praxis

Truthful and open communication is important for building trust, and for creating integrative interventions (see Chapter 6: NEGOTIATION). Transparent, collaborative monitoring of program performance, together with a sense of ownership, also makes the adaptive improvement of conservation interventions much easier (see Chapter 6: NEGOTIATION and Chapter 8: RESPONSIVESS).

As mentioned earlier, declaring and periodically reminding the community about our main purpose is a first step in transparency. Incidentally, it also helps in reiterating the conservation requirements and linkages of any community-based intervention, something that is not always strongly conspicuous in the interventions, or easily retainable in peoples' memories (see Chapter 6: NEGOTIATION).

It is useful to combine the disclosure of our main purpose of conservation with reiterating our desire for collaborative efforts that ensure beneficence and non-malfeasance for the community. This is especially helpful in putting people at ease, particularly during the initial interactions with communities.

It is useful to combine the disclosure of our main purpose of conservation with reiterating our desire for collaborative efforts that ensure beneficence and non-malfeasance for the community. This is especially helpful in putting people at ease, particularly during the initial interactions with communities.

Another aspect of transparency is the clarity of shared objectives, norms and rules of any intervention, and the roles and responsibilities of the conservation organization, the community, and individual members of the community. This is discussed in detail later (Chapter 6: NEGOTIATION). For now, it is useful to keep in mind that transparency does not imply making a single set of rules and trying to implement it uniformly. As discussed earlier, every initiative is ideally tailored at the level of the community, and it is at the individual community level that there is clear identification of roles and responsibilities.

Transparency in choice

Sometimes, choices need to be made from within the community. For instance, while one tries to cover as many of the households as possible in community-based conservation interventions, there are situations when maximizing participation is not desirable: e.g. while trying out a new intervention, when it is prudent to work with a small but representative sample from within the community.

How do we decide whom to work with? Preferably, we don't, and it is the community that makes such choices collectively instead. This helps prevent inadvertent factional alliances, and it helps rule out perceptions of favoritism.

Local communities commonly face such issues of choice. These come in the form of occasional employment from the government or other livelihood opportunities to the community where the demand outstrips supply. Or they come in the form of responsibilities that a few of the households are asked to shoulder on behalf of the entire community.

Most local communities have fair and transparent systems to deal with this. They work on rotation or through drawing lots between households. In situations where individual skills or characteristics required by our experimental design or other program needs are deemed necessary, communities can also potentially adopt mechanisms for incorporating qualifying benchmark skills or conditions into the selection system.

Where such systems are available, it is best to explain the requirement to the community, and rely on their system to make the choices amongst households or individuals. In cases where the suitability, transparency or equitability of such local systems is in doubt, it is still important to involve the community, assist in framing the rules of selection, and to respectfully ensure that those rules are followed in making the choices.

The same principle applies to the hiring of people. We often hire individuals from the local community to assist with research and conservation work. Some of them, in fact, have grown over time into becoming highly effective conservationists in their own right.

More often than not, we make the choice of which community members to hire. Though we may seek the advice of community leaders, elders, or generally knowledgeable people, the selection and choice is the conservationist's. At least, this is how I worked initially. When the program was relatively small, it seemed to work. When it started growing larger, there were problems.

In some communities, we were no longer allowed to do this. They had rather strong and equitable systems of distributing opportunities amongst the households, and insisted on us following them. The system worked on rotation, and, in some of them, the beneficiary could hold the privilege only for a specified period of time, say a year or two, after which, the job would shift to another household.

Not surprisingly, this wasn't easy to accommodate. Not every individual has the same interest, or is equally capable. And the lack of continuity meant that constant effort had to be invested in orienting the person to the job requirement.

Our system wasn't working. But nor was the community's. Community leaders were pointing to the unfairness to the households when we made the choice. I was pointing to the problems we faced when they were choosing. Finally, through discussions and negotiation, we worked out an integrative solution (see Chapter 6: NEGOTIATION).

In the new system, the communities still make the choice, and they do so transparently and equitably. However, their final choice is from a truncated pool of households that have potential candidates specifically suited for the job at hand, based on the requirements we provide them. These usually include standards of competency in language and communication skills, level of education, and the extent of travel and fieldwork likely to be involved. The community doesn't insist on any time period to be imposed on the terms of hiring. If a particular hire doesn't work out, we go back to the community representatives to discuss the issue. The problem is resolved amicably through a new hire, following a similar procedure where the community plays an important and transparent role.

While removing a program staff member belonging to the community, it is similarly useful to ensure that the community representatives are fully involved in a perceptibly fair and respectful process of removal. Helping protect the individual's dignity, if not their job, becomes even more important because, unlike in a regular job, the individual does not go away after removal, but continues to live in the same community.

Providing the person a chance to explain, and if removal is mandated, encouraging them to resign rather than be removed, allows a face-saving avenue. Unless the problem is due to some serious and unacceptable misconduct, it is best to rest the blame on circumstances or a mismatch of needs and skills rather than the individual's incompetence. Otherwise, a seemingly small issue like a staff change can potentially have a disproportionate effect on the larger relationship with the community.

> The potential positive influence of local champions on the community for conservation or other pursuits tends to erode when financial rewards gets involved, even if entirely legitimate.

There is the related issue of what I have earlier referred to as local champions. It is useful to recognize that, more often than not, behind the successful implementation of conservation interventions at the community level is usually the disproportionate influence of one or more individuals from the community.

There is often the temptation to hire such individuals. This is certainly a convenient short-term arrangement, might sometimes be useful, but is not always a good idea. The potential positive influence of local champions on the community for conservation or other pursuits tends to erode when financial rewards get involved, even if they're entirely legitimate.

Fortunately, these local champions tend to also be amongst the more self-sufficient members of a community, and are often less in need of livelihood opportunities. Like Makhan (see Chapter 1: PRESENCE), they are often motivated by the pride of being involved in programs of societal relevance, and by their relationship with the conservationist, rather than by a desire for personal gain.

Dos:

- Disclosing our purpose and clearly communicating conservation goals to the community
- Reiterating our desire for beneficence and non-malfeasance
- Maintaining transparency whenever making choices, such as the selection of households for a pilot intervention, or hiring of community members as program staff
- Interacting periodically with a broad set of community members, not just leaders or local program coordinators

Don'ts:

- Withholding information from communities, especially about potential negative impacts of conservation interventions
- Making decisions and choices without consulting the community
- Hiring local champions as paid program staff

Chapter 6:
NEGOTIATION

Community-based conservation partnerships require negotiations for arriving at robust joint agreements and for increasing ownership. However, unlike market transactions, these are not negotiations where bargaining and shrewdness lead to the best results. Personal relationships facilitate effective negotiations, and developing mutual trust as well as discussing interventions individually with key community members should precede formal community negotiations.

The negotiations are best done in an integrative manner – rather than through positional bargaining – based on sharing of information and interests, use of objective standards, and building incentives and tangible stakes in the interventions for the community. Agreement, once reached, must be formalized in the form of signed working documents that record program details and the roles and responsibilities of various stakeholders. Unlike the market, where we have the option of shopping elsewhere if we don't like a deal, walking away is not an option in community-based conservation of landscape species such as the snow leopard. If negotiations do not progress, greater investment in communication and relationship building is recommended, as is third party mediation, a concept that local communities are usually familiar with.

Conservation goals often come into conflict with the goals and aspirations of local people, their livelihoods, traditional resource use, and their desire for development projects. Recognizing that this dichotomy is a shared problem, and that sustainable solutions can best emerge through engagement, dialogue, and negotiation is an important step in effective conservation and human development (Redpath et al. 2013).

Community-based conservation efforts focus on such shared problems. These include, for example, the desire to conserve biodiversity on land owned or used by people, or the desire to conserve species that are the cause of economic losses to the local people.

The problem may be shared, but it is a problem in the first place because there is almost never a perfect match between the interests and expectations of the community, and those of the conservationist. Effective negotiation, therefore, is central to community-based conservation.

Patience and persuasiveness

Before the actual negotiations on conservation interventions, however, it is important to create suitable conditions to begin them. This requires time and effort, especially if they are to effectively lead to biodiversity conservation, and especially for those interventions that rely relatively more on trust and social capital. It requires active listening. As discussed earlier, building relationships and establishing trust is critical (see Chapter 1: PRESENCE). It can only happen through sustained interaction.

Conservationists are an impatient lot, especially when young. I was no exception. Many years ago, I had set out on a trip to initiate community-based conservation efforts in a remote Himalayan site. Having completed some field surveys and done a threats assessment, the following year I decided to visit some of the communities with the idea of starting a conservation partnership.

I remember sitting around a fire and talking with the head and two local leaders of one of the larger communities in the region. I was telling them about our desire to start a conservation partnership. I was viewing this as an initial introductory interaction, which I had hoped would be followed by a meeting with the representatives of all households in the village, which in turn would eventually lead to the conservation partnership.

This group of local leaders, meeting me for the first time, missed several nuances that I had tried to prematurely and hurriedly communicate regarding our potential long-term conservation partnership. But they clearly understood the main purpose of my visit: that I wanted to start a livestock insurance program.

I have to admit that I did introduce them to the livestock insurance program that communities in other regions with similar problems were running in partnership with us. However, not even once in the course of the conversation had I directly asked them whether or not they were interested in starting one.

Their interpretation of my intent, however, wasn't inaccurate.

Even before the trip, based on the earlier knowledge we had gathered, I had decided that a community-based insurance program would be useful for them and the carnivores in the region. The intervention had, after all, helped in other areas faced with similar issues. And indeed, I had made it my personal goal for the trip to start this intervention in one or two communities.

Our conversation ended abruptly, as did the plans for a conservation partnership. There was no community meeting. The leaders thanked me for visiting them, and said they were not interested in starting a livestock insurance program.

I had just reached this village a couple hours before, after traveling for three days by public transport and another day on foot.

Thinking back, almost every step I took was in contradiction to what I have written in this document. I was too focused on the intervention, and I had already decided on my own what it would be. Rather than concentrating on communication, I was in a hurry to start the intervention, before even beginning to build a relationship with the community, let alone trust.

How often have we turned away sales representatives arriving unannounced at our door? Including insurance sellers. Why shouldn't we expect communities to do the same?

In community interactions, when we say things matters as much as what we say and how we say them. Therefore, striking the right balance is important.

While conservation threats are usually urgent and require rapid action, pushing for urgent decisions or action is usually a deal-breaker in community-based efforts. In community interactions, when we say things matters as much as what we say, and how we say them. Therefore, striking the right balance is important.

Moving away from positional bargaining

Negotiation is a pervasive fact of life and everyone engages in it most of the time. The most common and obvious form of negotiation involves haggling over the price of something, and is called positional bargaining (Fisher et al. 1991). Both parties state and defend their position, usually starting from relatively extreme opposing points, and discussions lead to potentially finding a mutually acceptable solution.

This is perhaps an appropriate negotiation strategy for one-time interactions, where the parties do not need to engage any further. Although commonly employed, it is only partly useful for specific issues like price negotiation, where one party's loss is another party's gain.

Because it is such a common form of negotiation, one occasionally faces or ends up engaging in positional bargaining attempts in community engagement, especially initially when the relationship with the community is new. Bargaining might take place, for instance, over the amount of funding to be received from the conservation organization to strengthen the insurance corpus, or over the purchase price of handicrafts in Snow Leopard Enterprises, or even in general negotiations.

There was an interesting experience in our first visit to a village in an important, relatively remote Trans-Himalayan valley. This village lies near important snow leopard habitat. Because this was the first visit, it was aimed only at making initial contact with the community, and to introduce our work.

After the initial introductions and discussions during the day, we had nearly ten community representatives visiting us in the evening, in one of the village houses where we were staying with a family. After politely thanking us for our interest in engaging with them, they placed an unexpected demand.

We were asked to hire 3-4 young people from the village. We were told frankly, albeit with utmost politeness, that should we not be in a position to employ their youth, we were not welcome to work in their area.

Their demand, which was made within a few hours of us reaching their village for the first time, was uncharacteristic and surprising. It was evident that it was an attempt at positional bargaining. Presumably, their hope was that in the negotiations that would follow, we would mutually agree to support at least one person's employment. Indeed, I had just been talking to them a lot about the biodiversity importance of their area, and the need to work there, which perhaps to some extent encouraged them to take this stand.

Although surprising, their position was understandable. Employment opportunities are scarce in this remote village located in a faraway side valley, compared to some of the other villages where tourism has begun to flourish, alongside other external livelihood sources. The idea of employment itself may have been motivated by the fact that our team included staff members and temporary assistants from another village where we had been working for many years.

It is also possible that others like us, interested in collaborative work of some kind, may have visited the village, and, deliberately or inadvertently, created expectations in the community that remained unmet. Through positional bargaining, the community was perhaps trying to secure at least some tangible gain – or at least a commitment – from our interest in their area. Who knows if we would otherwise ever return?

One is occasionally confronted with such situations in community engagement.

Positional bargaining can be inefficient, and has the potential to harm the relationship between communities and conservationists. Because effective positional bargaining involves withholding information, it is not transparent, and therefore ethically unsuited for community-based work.

How does one respond when faced with such positional bargaining during community engagement? By changing the game, in the words of Fisher et al. (1991).

Towards integrative solutions

While positional bargaining involves distributive solutions, and is best served by withholding critical information, enhanced communication can help create integrative solutions (Smutko 2005). By expanding the scope of the initial bargaining and creating joint value, integrative strategies aim for mutual gain. They rely on sharing of information, truthful and open communication, and focus on the actual interests of the parties rather than their positions (Fisher et al. 1991, Smutko 2005).

Let's get back to the Trans-Himalayan village, where we sat with a group of community representatives, facing the ultimatum of providing employment to village youth or being banished from the area. We had three choices. We could refuse to negotiate, walk away, and find other communities to work with. We could agree to their terms or bargain for fewer staff positions. Or we could take an integrative approach and change the terms of negotiation.

After hearing their demand patiently, we engaged the community representatives further, trying to understand their interests better, and to communicate our interests more clearly.

We expressed that we understood their concerns, especially regarding the inadequacy of existing employment opportunities for their youth. We reiterated that our main interest was to conserve wildlife with their support. We were upfront in communicating that though we understood their concerns, our relationship with them would be a non-starter if they were to view us as a source of employment, even if some of our programs could help with livelihood enhancement over the longer term.

We were to leave the next morning. I requested the representatives to explain our interests to the entire community once again the next time they all met. I reiterated our desire to build a long-term relationship with them, and indicated that irrespective of the position they had taken, we would continue to visit and interact with them. It was late evening, the meeting concluded with the representatives assuring us they would communicate our views to the entire community, and we said our goodbyes.

As surprising as their demand for employment had been, even more unexpected was what followed after they left. As we prepared to curl up inside our sleeping bags for the night, barely half an hour later, all the community representatives streamed back into the room again, some of them betraying hints of embarrassment on their faces.

They were back with a new message for us. The message was that the community understood our point of view, and that they were retracting their demand for employment. They would not be imposing any conditionality on our work. They requested us to come back and begin working in their area.

Our team went back. We started by monitoring wild ungulates and camera trapping snow leopards in the area for three years. Our engagement then grew to include predator-proofing of the community's corrals, and the creation of the largest village reserve amongst all our partner communities, protecting c. 240 sq. km. of prime snow leopard habitat. Six young people from the village are getting trained in wildlife monitoring. Two of these six youth serve as paid village reserve guards at any given time, the positions rotating amongst all six of them. The community is now discussing with our team the possibility of initiating a livestock insurance program, an intervention that relies relatively heavily on trust and social capital (Chapter 11).

On our first ever evening in this village several years back, people had come at us with a hard positional bargaining stance. We were faced with three negotiation choices. We could have walked away from the village, and lost a huge opportunity for conservation. We could have agreed to their premature demand for employment. While that would have helped a couple of their youth, it would have forever created a mismatch of expectation between the community and us. It would have based our relationship on pressure and positional bargaining, rather than on empathy, trust, and cooperation. We took the third path. Honest and integrative negotiations served conservation well, as they did the community and ourselves.

That first evening, years ago, I had expected the community to change its position, but only over time, through further discussions during subsequent trips. An almost instant turnaround, though, was completely unexpected. Such can be the power of truthful discussion and an integrative approach. In community-based conservation, one is served well by assuming that most people one has to interact with are fundamentally decent, and fundamentally smart.

The sense of ownership over a program comes not just from being responsible for running and managing interventions, but also, importantly, from the people's role in actually designing or tailoring the interventions.

Negotiation and ownership

The resilience of conservation partnerships and interventions relies heavily on the extent of ownership people feel over the program. The sense of ownership over a program comes not just from being responsible for running and managing interventions, but also, importantly, from the people's role in actually designing or tailoring the interventions.

In one instance, two small neighboring communities jointly expressed an interest in participating in the livestock insurance program. Because they were too small individually or collectively for an insurance system to become self-sustaining, we helped facilitate their participation in the ongoing insurance program of a nearby larger village.

This seemed to work well, until, during one particular year, these two communities faced relatively high livestock losses to wild carnivores. As most of the livestock was insured, the owners were able to get compensated. Once they received their compensation amounts, however, members of both the communities stopped paying further premium, effectively exiting the program.

What went wrong here? Although it is difficult to know for sure, looking back, it is clear that the members of these two communities felt little sense of ownership over the program, which was seen as belonging to the larger village. They had joined a pre-existing program. There had been no negotiations with them, and no mechanisms followed to seek their inputs into designing the program. In the absence of negotiations, and, therefore, ownership, these community members were, perhaps understandably, predisposed to maximizing immediate return while discounting the future costs and benefits. Although they decided to walk away from the insurance program, we obviously did not walk away from them. Our interactions continued, and after a few years of engagement, they rejoined the program.

Contrast their approach with that of the participants from the larger village, whose program they had joined. At one point, the insurance committee members, who change on rotation, themselves raised a concern regarding the erosion of their insurance corpus. They realized that a lot of funds were being used up to compensate the carnivore-caused mortality of yak calves. However, because the premium amounts till that point had been based solely on the market value of livestock, and not on their risk of mortality, the amount being contributed by people for insuring yak calves was relatively low.

The community readily agreed to fix this mismatch between premium and compensation rates, mediated by the risk of mortality, by increasing their premium contribution for yak calves by some five times. Such is the value of building ownership of the community over any program.

Fair standards

Fisher et al. (1991) underscore the importance of using objective criteria when negotiating the terms of any agreement. In our case, we may again use the example of deciding the compensation amounts in insurance programs according to the market value of livestock while at the same time correcting the premium amounts people contribute based on the risk of mortality.

Fair standards, apart from market value, can also be based on expert opinion, laws, or customs, or a combination of criteria. For instance, the purchase price of handicrafts produced by women involved in Snow Leopard Enterprises is negotiated based on a combination of raw material, time and skilled labor involved in each product, and the market value (Chapter 10).

Not following objective criteria can lead to problems in the future. When we set up the first village reserve, we capitalized on the fact that the local community had been traditionally leasing out their land for a grazing fee to migratory pastoralists, but had discontinued the practice due to fears of land degradation. While negotiating the terms for the village reserve, we used the payment they used to receive from the migratory pastoralists as a benchmark.

We failed to follow any objective benchmark while negotiating with another community for one of the village reserves we subsequently established. There was a sense of urgency while setting-up this reserve, dictated by the need to help arrest the imminent local extinction of the Tibetan gazelle *Procapra picticaudata*. We ended up in a situation where, for a small but critical area to be freed by the community from livestock grazing and other forms of resource use, we agreed to a relatively large amount to be paid annually as offset.

This was many years ago, but even today, the problem hasn't fully gone away. After a recent round of negotiation, the community has tentatively agreed with our suggestion to link the protection of this reserve with a new livelihood-linked conservation initiative called Snow Leopard Friendly Cashmere that we are piloting.

Respectful negotiation

Following rational approaches and fair standards in community-based negotiations is valuable. However, the importance of relationships and respect in negotiation is not to be underestimated at any cost, and must be reiterated. Perhaps it is again best done with an example.

The agreement with the community pertaining to our first ever village reserve, which had been expanded to three times the original size during the first renegotiation after the initial five years, had come up for extension for the third time. The reserve had been in existence already for a remarkable 15 years, and we were to renegotiate the terms of extension for another five years.

It came as a surprise, considering our long association and partnership with this particular village, when I learnt over phone that the negotiations with the community, being conducted by some of our team members, had failed. It wasn't clear why, but based on a request that I passed on over the phone, the community representatives agreed to defer their final decision till I arrived.

A few months later, I reached the field and met the community representatives. Surprisingly, this time, the discussions proceeded without any hitches, and the agreement to renew the village reserve for another five years was reached smoothly, over casual conversation and cups of tea.

> Many local communities, and indeed, cultures, view people and relationships as being central to negotiations and partnerships. While communities will obviously be inclined to be involved in interventions that benefit them, it isn't just a mercenary cost-benefit analysis that motivates people.

It turned out that during the earlier round of discussions, our team members, at some point during the negotiation, had ended up communicating that if the community was not interested in renewing the agreement, there were always other options available to us; other communities in the region that we could choose to work with. The community representatives felt disrespected, and the negotiations were suspended.

The village reserve, by helping this community access funds for community development, had been – and continues to be – viewed as being beneficial. But the disrespect that the community representatives presumably perceived because of a few words that were used during the negotiation was enough for them to nearly call off the agreement. The words, clearly, seemed to have been used out of impatience, or used as a negotiation tactic, neither being contextually appropriate.

Many local communities, and indeed cultures, view people and relationships as being central to negotiations and partnerships. While communities will obviously be inclined to be involved in interventions that benefit them, it isn't just a mercenary cost-benefit analysis that motivates people. The role of relationships and respectful communication is not insignificant.

Communication and third party involvement

We tend to view the implementation of interventions as success, and not being able to do so as failure. Such an intervention-focused, win-lose approach with communities usually comes at the cost of inadequate investment in the relationship, and is counter-productive for conservation.

Relationship building is not about winning or losing. Sustained communication with the community collectively, and discussing motivations and proposals individually with community members are rather useful in aligning peoples' thoughts for generating effective conservation interventions.

It is invariably helpful to discuss the intervention ideas individually with key community members, before making formal proposals and initiating negotiations with the entire community. I have also found it useful to seek out and discuss the ideas individually with people who are, for some reason or the other, expected not to be supportive.

These steps, especially when taken before formal community meetings, help to get insights regarding the kind of concern and opposition one might face, and to think through ways to address them, thereby better preparing the conservationist for the negotiations. Some of the ideas obtained in this way help make the intervention more apt. Such discussions also help in generating support for the idea and promoting ownership, especially amongst people whose inputs have been sought and considered in advance.

After barely a few months of immersion as a researcher, when I first broached the idea of setting up a village reserve – our first one discussed above – with the community, there was understandable skepticism among village elders. They had concerns about their land. Could getting into an agreement with us be the first step in losing rights over their land, and for it to be eventually acquired by the Government for wildlife conservation? Given the global history of coercive conservation, it wasn't an unfounded concern.

I had happened to discuss the idea with some of the village youth earlier, not as a negotiation tactic, but simply because seeking their inputs was a natural thing to do. I used to interact with them much more, join them in volleyball and cricket contests, and occasional impromptu contests over consumption of locally brewed alcohol.

As the discussions with the elders unfolded in my first-ever formal community meeting, something fascinating happened. As the elders voiced their rather legitimate concerns, unexpectedly and without any prompting, many of the youth decided to take upon themselves the challenge of convincing their elders. I distinctly remember largely becoming a mesmerized spectator, speaking only occasionally when clarifications were sought. The negotiations soon culminated in an agreement.

However, there will be situations when the negotiations do not move forward despite all the effort, and patient and respectful communication. Under such situations, third-party mediation is recommended (Fisher et al. 1991).

If there is a neutral mediator that community members can trust, such as a respected member from another community in the same region, it can to some extent help circumvent any trust deficit that might exist between the community and the conservationist. A good mediator is able to understand and help better communicate the interests of both the parties to each other, and assess any hidden fears and concerns more accurately.

Third party mediation is not an alien concept amongst local communities. It is commonly employed in negotiation and conflict resolution within and between communities, which are often mediated by a group of neutral elders. When conservation negotiations with the community reach a stalemate, this is a useful option to consider.

BATNA and when to walk away

Negotiation strategists underscore the importance of thinking through what they call the Best Alternative to a Negotiated Agreement in any negotiation (Fisher et al. 1991). BATNA is the course of unilateral action in case there is no joint agreement. More importantly, it is a useful benchmark against which to compare the cost-benefits of any potential joint agreement. Doing so prevents one from accepting terms that may be too unfavorable, or from rejecting terms that may be useful to accept.

In standard negotiation strategy, if the best potential agreement is poorer than the BATNA, walking away makes sense. However, when it comes to community-based conservation, even though we do often tend to walk away, it is not a straightforward and rational calculation, or a desirable option, as I will explain.

Our typical response when discussions with a community seem to indicate a lack of interest, or when there is lack of progress in being able to implement a conservation intervention, is the walk away alternative. The decision is often guided by our desire for efficiency, and the reality that there are usually more communities than one could possibly work with, and that many of them are much more forthcoming. There is also the pressure of having committed to funding agencies that interventions would be implemented in a time-bound manner.

As a general rule of thumb for landscape level-conservation, however, walking away is problematic, and it must be considered a least favored or non-existent alternative. While we might walk away, the threats to conservation in a landscape, or the threats to a landscape species, do not. If the threats are internal and emerge from the community, walking away can actually intensify the problem.

In landscape level-conservation, walking away is problematic, and it must be considered a least favored or non-existent alternative. While we might walk away, the threats to conservation in a landscape, or the threats to a landscape species, do not. If the threats are internal and emerge from the community, walking away can actually intensify the problem.

Indeed, it is prudent to concentrate the interventions in situations where the community involved is forthcoming in establishing conservation partnerships. However, that should not come at the cost of neglecting communities that are less inclined to form conservation partnerships. From a conservation perspective, it may be important to invest more time, rather than less, in such communities. Instead of interventions, the main investment in such situations needs to be on communication and relationship building.

Walking away, or quitting, therefore, is really not an option in community-based conservation of landscape species such as the snow leopard.

Written agreements

Community-based conservation must be approached in an à la carte fashion. Innovation, site specificity, and flexibility are valuable. It is important that one does not feel bound by rules while developing creative interventions with the communities. Both time and flexibility are required for accommodating the constraints and using the opportunities, for addressing the course of events, for building trust, and for designing contextually appropriate and creative interventions.

Flexibility, however, does not mean a lack of order. When there is broad agreement on the need and scope of any intervention, clear identification and distribution of responsibilities and regulations is essential. Written agreements must be employed for recording them, and can help address a variety of issues. They can especially help in keeping in focus the important nuances of any program that may be easy to forget or overlook.

As in any other relationship, misunderstandings and a mismatch of expectations is often a problem in community-based conservation, despite constant communication. Many participants initially thought, for instance, that in the livestock insurance program, they would get back their entire premium contributions if they did not lose any livestock.

Recording the details of any intervention in the form of a signed agreement has a valuable role in helping both the community and the conservationists develop a similar understanding of program details, as well as of their respective responsibilities and privileges.

Agreements also play a role in helping bring more equity among the participants. We have faced situations, for instance, where, in Snow Leopard Enterprises (Chapter 10), the distribution of orders wasn't equitable, and instead, became biased due to favoritism of the local coordinators. Sharing a copy of the agreements and contracts with all households, or additionally signing individual contracts with each participant, can help address such problems, and bring more equity and transparency (see Chapter 5: TRANSPARENCY).

Interventions usually aim at changing behavior and garnering community support for conservation. A formal set of interventions, recorded in the agreement, helps formalize the system to positively influence conservation behavior and community welfare.

Social dilemmas, or the conflicts between individual and collective interests (Karp, 1996), tend to influence both the conservation related behavior of individuals and the outcome of community-based efforts. Building sanctions and incentives into the intervention can help encourage conservation-friendly behavior and address social dilemmas. This is best done through community discussions, and recording mutually acceptable clauses into the signed agreement. As a rule of thumb, because of their positive connotation, incentives for conservation-friendly behavior are to be preferred over sanctions for non-compliance.

When participants have a stake in the proper running of an intervention, it is more likely to run well. Cost or effort sharing is one of the ways to increase this stake. Because people also contribute premiums to build the insurance fund, cases of false claim attempts are few, far between, and not tolerated. On the other hand, preventing fraud would be a major concern were it a compensation program set up entirely with Government, NGO or private funds.

In Snow Leopard Enterprises, communities commit to preventing poaching in their area, and ensuring that the commitment is honored enables all participants to get a bonus in addition to the purchase price of the products they produce (Chapter 10).

Generally, when a livestock predator such as a snow leopard or a wolf gets killed, one cannot really expect herders to complain about it. In their view it may even be a cause for quiet celebration, as it means one less "enemy" to deal with. In SLE, however, should any instance of poaching of a snow leopard or prey be detected, all participants lose their bonus, thus creating a positive incentive – rather than punitive action – and peer pressure towards conservation-friendly behavior.

When one of our radio-collared snow leopards in the Gobi-Altai was trapped and killed by a herder, the community was upset. The herder happened to belong to a community involved in Snow Leopard Enterprises. The SLE participants insisted to the local governor that in addition to the herder being made to face the legal consequences of his action, he must be made to pay the bonus SLE amount that the community had lost due to his violation of the community's conservation contract.

SLE thus ensures that people have a financial stake in conserving. Agreements that ensure tangible stakes for the community in the program process and impact combine powerfully with the sense of ownership and pride to strengthen conservation programs.

An agreement – best written in a positive tone – should clearly identify the roles and responsibilities of the parties, suggest the course of action in cases of violation, and state that it has been arrived at as a result of discussion and negotiation between the parties. In the absence of such agreements, the interventions are unlikely to have the desired impact on biodiversity. Further, the resilience of both the intervention and the relationship with the community can especially be threatened if there is no pre-agreed mechanism to respond to breaches and instances of conservation-unfriendly behavior that the program is designed to address.

The resilience of both the intervention and the relationship with the community can especially be threatened if there is no pre-agreed mechanism to respond to breaches and instances of conservation unfriendly behavior that the program is designed to address.

Sometimes, serious offences get

committed; the law of the land is broken, such as when the snow leopard was killed by the herder. This can create a dilemma for the conservationist, especially if the offence is committed inadvertently.

If a law enforcement agency, such as the Forest Department, happens to be one of the partners in the community-based effort, this is easier to handle. If that is not the case, the conservationist feels compelled to report the instance to the agencies, but that can come at the heavy cost of losing the relationship with the entire community. Under such circumstances, detailed discussion with the community regarding the incident and the course of action must precede the lodging of formal complaints.

During general negotiations and drafting of agreements, it is also helpful to convey to the community that the agreements are not carved in stone, and that, in fact, it is useful to modify the agreement with mutual consent as we learn from our experiences and mistakes to jointly and adaptively improve the interventions.

Apart from its practical value in facilitating adaptive improvement, projecting the agreement as a working document also helps in putting the community at ease from any concerns arising out of entering into an agreement whose full consequences they might not immediately understand.

Agreements generally help to ensure that both the conservationists and the community jointly take responsibility, especially when things go wrong. This implies jointly investigating any breaches of agreement, not necessarily with the idea of finding individuals and sides to blame, but finding solutions.

Problems are potentially valuable opportunities to improve community-based programs, as we shall see later (Chapter 8: RESPONSIVENESS). Agreements provide a useful platform for such improvement.

Dos:

- Engaging in integrative negotiations with communities that focus on interests rather than positions
- Employing transparent, objective criteria or fair standards in negotiations with communities
- Bringing third-party mediation if negotiations aren't moving forward
- Discussing potential conservation interventions individually with

community members before formal negotiations with the entire community
- Involving community members in the design of interventions
- Recording details and nuances of community-based interventions through written agreements
- Including mechanisms that allow for revisiting and making changes to signed agreements
- Building in incentives and tangible stakes against social dilemmas or violation of conservation agreements

Don'ts:
- Engaging in positional bargaining for a bigger piece of the pie
- Pushing the community to make urgent decisions
- Withholding information
- Walking away from the community if negotiations aren't moving forward

Chapter 7:
EMPATHY

The importance of empathy in community-based conservation cannot be overstated. Empathy enables the practitioner to assess the idea and the costs of conservation and conservation interventions from the perspective of the local people. It helps understand that while conservation might be the foremost pursuit of the practitioner, it can sometimes be but a minor concern for a community member who is dealing with economic hardship and other issues. Empathy allows the practitioner to be more accommodating towards local people and more appreciative of their conservation effort. When things go wrong, it helps focus on the root causes rather than on individuals or perpetrators. Empathy is a skill that can be enhanced with practice and through immersion in the community.

Empathy, which involves the perception and understanding of the ideas and emotional state of others, is rather important for effective community-based conservation, as it presumably is for any social work practice (Gerdes and Segal 2011).

Empathy enables sensitivity to other people and cultures, and a better ability to view the problems and the opportunities from their perspective. Empathy therefore helps the practitioner assess the idea of conservation and conservation interventions from the perspective of the community. Similarly, our ability to identify conservation opportunities or to create meaningful conservation interventions is influenced by our understanding and empathy.

It was our awareness of a community's past practice of leasing out relatively distant pastures to migratory herders that gave birth to the idea of leasing out land for wildlife recovery. The community agreed to the village reserve because the idea wasn't alien to them. Leasing out land was a familiar concept, and in this case, their pastures would get much needed rest from grazing, rather than being degraded as in the earlier system due to intensive grazing by migratory livestock. In the absence of familiarity and empathy, it would have been difficult for us to conceive the idea of village reserves.

Research suggests that empathy is not just an inherent quality, but can be taught and increased, thereby making it possible for conservation practitioners to become more skillful and effective (Gerdes and Segal, 2011).

Immersion in the community is an important way to improve contextual familiarity and our ability to empathize with them (Chapter 2: PRESENCE). A simple but important lesson one learns through such immersion is that while conservation may be the most important pursuit for us, it is only one of the various aspects of life, sometimes even a peripheral one, for the local people.

> A simple but important lesson one learns through such immersion is that while conservation may be the most important pursuit for us, it is only one of the various aspects of life, sometimes even a peripheral one, for the local people.

This might appear like a small issue, but is actually an important lesson for any conservationist. At a minimum, it teaches us to be patient when, for instance, people are unable to gather for community meetings as planned. It helps us appreciate the community's time and support for conservation programs much more than we would otherwise do. It helps us better gauge what kind of conservation interventions would be effective in a given situation.

And in many ways, it makes our attitude towards our conservation partners – the local communities – more accommodating, generous, and understanding. It teaches us patience, for instance when people are sometimes not able to honor their conservation commitment. When conservation agreements get violated, empathy helps us try to identify and address the root causes and improve the program, rather than only looking to apportion blame and fix responsibility. Blaming, incidentally, is consistently counter-productive in such partnerships, even when it is justified (Fisher et al., 1991).

Livestock predation by large carnivores is often due to human error, and in such cases, it is natural to rest the blame on the herder. While grazing in the mountains, some livestock get separated from the herd, and it is these stragglers that tend to get killed more by predators. We know this, and the local people know it. Yet, livestock predation happens often – partly because herding in the mountains is not an easy job, partly due to lax herding, because the herder was drunk, or unwell, or whatever.

In many ways then, livestock depredation in the pastures is really the local community's problem; why can't they just herd better? This is a legitimate point of view. But it doesn't capture the whole picture, as some amount of empathy and common sense can help us understand.

How often do we end up partying late into the night, knowing fully well that we should have been at our desk, completing that important and much delayed, unfinished assignment? Or we take the car out for a short drive to the nearby store, knowing fully well that the insurance has expired and the car should not be on the road because it is both dangerous and illegal? We are usually aware of what the 'right' thing to do is, but we don't do it. We are only human.

I had earlier mentioned the incident where a snow leopard had entered a corral and killed almost the entire livestock herd, and in turn had been killed by the local people (Chapter 4: Respect). By the time I managed to visit this hamlet, some two years had passed. When we reached the house of the herder involved in the incident, a badly injured cattle and a dzomo (female yak-cattle hybrid) stood outside, receiving basic veterinary care.

They had been attacked by a snow leopard the previous day in a gorge near the hamlet, but had managed to escape with the injuries for which they were being treated. This hamlet is located right inside snow leopard habitat, and occasional instances of livestock predation were to be expected.

After spending time with the herder, we visited the corral where two years prior, the incident of multiple killings had taken place. It didn't appear as if any significant effort had been put into predator-proofing of the corral, despite the catastrophic loss of livestock two years earlier.

This one was an old corral. Later, we visited the three other corrals in the hamlet, some of them relatively new; and found that not a single one of them was predator-proof. There were gaping holes to allow for light, through which a snow leopard could easily enter. Where there were windows, they were flimsy glass ones, some already broken.

Even as we examined the corrals, a quick and discrete discussion with field staff suggested that the cost of predator-proofing these already roofed, enclosed corrals would have been rather low, though it would have required some effort in material fabrication and transport from the welder at the nearest township. When I asked the owners why they hadn't made the corrals predator-proof even after the devastating incident, I did not get a satisfactory answer. Local people too, after all, are only human.

Our offer of collaboratively predator-proofing the corrals by jointly developing a plan and sharing the costs equally between the community and us was readily accepted. Within a month, thanks to the excellent follow-up by our field coordinator, Tanzin Thinley, and the cooperation of the local people, all corrals were predator-proofed. It cost us a total of US$ 125, and the community invested a similar amount. That is all it took.

Indeed, this is something that the community members would have been fully capable of doing on their own, both in terms of the effort and the cost. Yet, in this case, they didn't, even though it was really their problem. A conservationist, in such a situation, could easily decide to leave things as they were. After all, it was their problem, and the solution was perfectly within their reach, only if they made a small effort.

The conservationist could be forgiven for deciding to not get involved. But that wouldn't help anyone, including snow leopards. Indeed, a chance to establish a conservation partnership with a community would also be lost.

Empathy allows us to better understand why things get done in a particular way – or why they don't. Empathy helps us realize that sometimes, a little push and support is all that the community needs, just like we do at times. And it is up to the conservationist to play the role of the catalyst.

Another dilemma that the conservationist occasionally faces comes in the form of requests for obvious and dire needs of the community that are unrelated to conservation, but are nonetheless important. For instance, to help do something about the absence of basic educational or healthcare facilities in a community. Or to help a community get over catastrophic economic setbacks due to occasional weather extremes such as the dzud or a flash flood. As a conservationist, does one get involved in these issues at all? If so, to what extent? How does one decide?

While it is not sufficient, a high level of empathy is essential in being able to make informed decisions on such issues. But as I shall discuss in the next chapter, these are complex issues and difficult questions, with no easy answers.

Dos:

- Trying to look at conservation issues from the community's perspective
- Taking both rational and emotional aspects into account when making decisions
- Making the effort to increase our capability for empathy
- Assuming that most community members – like most other people – are decent and intelligent

Don'ts:

- Forgetting that our own behavior can often be irrational or irresponsible
- Walking away because of perceived inaction on part of the community, rather than catalyzing action

Chapter 8:
RESPONSIVENESS

Change is the only constant in community-based conservation. Practitioners need to be responsive to the changing threats to biodiversity, the changes within communities, and to the need for addressing weaknesses in conservation interventions. Monitoring programs therefore must accompany any community-based conservation effort. They need to include threats indicators to constantly evaluate the main threats to biodiversity, process indicators to evaluate how well the conservation interventions are being implemented, and impact indicators that help to assess the actual impact of conservation programs on biodiversity. Conservation is about identifying threats as well as opportunities, and responding to them promptly.

It is in the nature of responsiveness, however, that the practitioner will be faced with difficult decisions when it comes to important community needs that are unrelated to biodiversity conservation. Whether or not to get involved in such cases can be assessed by asking:

- How serious is the problem?
- Is the problem episodic or chronic?
- Do we have the expertise and the resources ourselves, or is it better to facilitate a specialized organization to help address the problem?
- Is it possible to assist the community to meet their biodiversity-unrelated needs through interventions that are biodiversity-linked?
- How mature is the conservation partnership with the community?
- To what extent does a biodiversity-unlinked intervention enhance the social capital?

As I mentioned earlier, timing is critical in conservation, and especially in community-based efforts. New threats to biodiversity are constantly arising, and timely and creative responsiveness is essential.

Responsiveness, however, is important not just because of new threats to biodiversity. There are problems, weaknesses and management issues in the interventions themselves that periodically arise or become evident, and need to be addressed. Or, the interventions may require audit and adaptive improvement, especially if monitoring suggests that the desired or expected conservation targets are not being achieved. However, we need to be responsive not just in addressing problems, but also in seizing conservation opportunities.

Problems as opportunities

Problems are to be expected to arise periodically – if not constantly – in any work, and especially in community-based conservation. They are no doubt a cause of frustration for the conservationist, but they should neither surprise nor distress us.

In many ways, problems are necessary for growth, and effective problem solving becomes both a measure and an outcome of conservation program resilience. Problems are opportunities to improve community-based conservation efforts.

In many ways, problems are necessary for growth, and effective problem solving becomes both a measure and an outcome of conservation program resilience. Problems are opportunities to improve community-based conservation efforts.

The first expansion of our very first village reserve became possible as a result of a problem. In our initial agreement, the community had agreed to protect half the section of a side valley from livestock grazing and other forms of resource use to allow for wildlife and rangeland recovery.

On one occasion, as I passed by the reserve area on my way back to the village after a long day's fieldwork, it was annoying to see several of the livestock that had been brought to a nearby area that day grazing inside the village reserve. The herders were not in sight.

Back in the village, I requested a meeting with community representatives. I had planned to remind them about their conservation commitment, with the hope that they would strictly instruct the herders to ensure that this was not repeated. I was younger and short on both patience and empathy.

The community representatives apologized for what had happened, but, unlike me, rather than appearing to be dissatisfied with the day's herders, their discussions focused on how the topography and boundary of the village reserve made it difficult for the herders to prevent livestock from straying in.

Then came the solution, their solution. Without changing any other term of agreement, they unilaterally decided to double the size of the village reserve. It was extended to cover the entire side valley instead of just half, making it easier for the herders to prevent livestock from straying in. They would also explain to the herders again the importance of keeping the livestock out of the reserve area.

I was delighted and educated; their decision meant that the problem had become an opportunity to protect more area for biodiversity. I was also humbled by the contrast in their way of thinking, especially their empathy, and the deficiency in mine.

One encounters occasional problems in the norms and processes of community-based interventions, and these too can become opportunities for strengthening both the interventions and our relationship with the community.

In the livestock insurance program (Chapter 11), the communities have a clause that if a livestock owning family decides to not join from the outset but wait and watch to see how the program works, they would have to pay a joining fee to start participating later. Otherwise, it would be unfair on the original participants whose premiums, along with our support, created the insurance corpus.

In a cluster of villages we have been partnering with in the Ladakh Trans-Himalaya, we faced a peculiar situation. Several 'wait and watch families' were interested in joining the program, which had been running well for several years. While the original participants welcomed their inclusion, they expected joining fees to be paid. However, the size of the joining fee that the new families were legitimately expected to pay was a deterrent for them to join the program.

At a subsequent meeting, as part of the negotiations to encourage the inclusion of new families and revise the premium and compensation rates (it had been almost eight years since these were fixed), we offered to assist the new families to be included. We suggested that we would pay the joining fee on behalf of the new families through conservation funds.

This gesture, costing less than US$ 25 per family, all of which would go into strengthening the insurance corpus further, was highly appreciated by the community. Because it was only a one-time offer, the insurance committee members voluntarily conducted a drive to make all the non-participating families aware of the offer of the joining fee being provided for. The participation in the insurance program went up by 25 %.

At times, the problems are rather serious, but they too can become opportunities for conservation. A few years back, on one of my first field trips to Mongolia, we realized how serious mining had become as a threat to the snow leopard habitats of the country, especially the Tost Mountains in the South Gobi Province. The Tost Mountains form the study area of the most comprehensive and successful snow leopard radio collaring project ever undertaken.

Our study area, we found, was covered almost entirely with mining licenses. As I mentioned earlier, although we had been running a community-based intervention for many years, the problem of pervasive mining had crept up almost unnoticed, or rather unacknowledged, from a snow leopard conservation perspective.

The situation was distressing. We had to reprioritize all our activities, and take up the threat of mining in Tost as one of our key preoccupations. Years of hectic efforts with the Central government began. Our team, led by country Program Director Bayarjargal (Bayara) Agvaantseren, worked closely with the local government, as the local communities too were opposed to the idea of losing their grazing land to mining. The fact that we had been running the Snow Leopard Enterprises program with local communities in the region helped us quickly get together and form an alliance to address the external threat posed by mining, both to the local way of life as well as the area's biodiversity. We did not have to start building a relationship with the local people from scratch. Little time was wasted.

Finally, this problem turned into a small opportunity when the government relented and agreed to the proposal of declaring this area as the 6500 sq. km. Tost Local Protected Area, which, at least on paper, created a large, connected landscape of wildlife reserves, with the Great Gobi National Park on the west, and the Gobi Gurvansaikhan National Park contiguous to the north.

The problem hadn't gone away though. Local Protected Areas offer only the weakest levels of protection under Mongolia's conservation laws. The status is temporary and time-bound. So our collaborative efforts with the local people and with the government continued, as we tried to acquire the status of a state-level reserve for Tost, which would be long-term, and afford a much stronger legal basis for protection.

It took many years of hard work, risks faced by Bayara's team members, and the tragic loss of our colleague Sumbee, one of Mongolia's young and rising conservation stars to whose memory this book is dedicated. Finally, in the summer of 2016, the Great Hural (Parliament) of Mongolia approved the proposal to turn Tost into a State Nature Reserve. Our work in the region, of course, hasn't ended. Conservation is a life-long job.

Tost is an illuminating example of a crisis being turned into an incredible opportunity for conservation. Tost is also a classic example of why community-based efforts, by themselves, are far from being enough to achieve sustainable conservation, unless supported by appropriate legal and policy frameworks, a subject I will come to later (see Chapter 8: STRATEGIC SUPPORT)

Responding to opportunities promptly

As I have discussed earlier, building resilient relationships with the local communities is most important; effective conservation interventions follow. At times, however, opportunities for conservation interventions present themselves when one least expects them, and when that happens, we must seize them; the relationship building then follows. Effective conservation is about timing, about creating or finding opportunities, and responding to them.

The predator-proofing of corrals in the hamlet where a snow leopard was killed is one such example (see Chapter 4: RESPECT and Chapter 6: EMPATHY). Our standard predator-proofing intervention is different from what we did here. In our standard intervention, we supply the raw material and inputs into design, while the community provides labor. Jointly, we strengthen the walls of the corrals, cover the ceiling with chain-link fence, and affix strong doors.

In this case, there were strong walls and doors in the corrals already. What was required was to reinforce the windows with predator-proof grills. We needed to act quickly to preclude the possibility of another predator attack. So even during our first visit, we suggested a cost-sharing arrangement for predator-proofing, and the community readily agreed. Within a month, all windows were reinforced.

Now that the corrals have been predator-proofed collaboratively, is our job in this community done? Far from it. Although we have a signed agreement with them where they have committed their support to conserving wildlife, our corral improvement effort should be viewed more as a community entry conservation activity; the beginning of a relationship. Constant communication must follow, as is now being ensured by our field staff.

The subject of community entry activity is a good point of departure to try and tackle a difficult question that conservationists are often confronted with. Should we, or rather, when should we get involved in community work that is unrelated to biodiversity conservation?

Responsiveness when societal needs are biodiversity-unrelated

I have visited a community of 300 or so in a tropical rainforest in the Eastern Himalayas, where more than 40 people had lost their lives during the previous two years to avoidable and treatable infections. They did not have access to basic healthcare.

I have had the privilege of breaking bread with some wonderful families in snow leopard habitats of the Pamirs, inside the comfort of their yurts, while it snowed outside. Their generosity notwithstanding, these people were food-secure for barely half the year, their lives ravaged by decades of war and factional fighting. They had to rely on aid and opium to get by during the rest of the year.

The desire to conserve gets us to difficult places and situations. How does one even begin to discuss conservation when people are at the edge of existence? Indeed, these are extreme examples, but they help to put the problem in perspective.

To rephrase the question, do we have an obligation to get involved in issues affecting the community that do not have anything directly to do with biodiversity conservation, our key mission? There is no easy answer to this question. Perhaps the answer is different in every situation, and for every individual or institution.

When people are living in abject poverty, livelihood-based interventions such as Snow Leopard Enterprises assume even greater importance. Livelihood enhancement can enable people to better deal with many of the challenges they face.

At least on the face of it, interventions like Snow Leopard Enterprises (Chapter 10) can potentially enable a win-win for conservation and human development. However, it is useful to be aware that unless the per capita livelihood enhancement is substantial, such efforts could deprive the already underserved people even further. For instance, in some situations, curtailing hunting can cut off a critical source of protein, which would be difficult to replace for the poor unless livelihoods are strengthened substantially.

> One useful way – though neither sufficient, nor perhaps always appropriate – to assess the extent to which we need to get involved in issues unlinked to biodiversity is to examine whether the problem and the need are chronic or episodic.

Interventions such as the livestock insurance program (Chapter 11) may also be useful for livestock-based people in poverty, though it would be important to keep premiums much lower, and rely on greater conservation subsidy to build the insurance corpus.

These interventions, of course, are linked to biodiversity. The reason to invoke them here is to reiterate the point that strengthening peoples' livelihoods substantially can enable them to better meet their basic needs of education, healthcare, and nutrition, which themselves may be less related to biodiversity.

Getting back to the main question of whether or not, as conservationists, we should get involved in addressing societal needs that are unrelated to biodiversity. One useful way – though neither sufficient, nor perhaps always appropriate – to assess the extent to which we should get involved in issues unlinked to biodiversity is to examine whether the problem and the need are chronic or episodic.

There was widespread devastation in China following an earthquake in 2010 that was epicentered near Yushu, Qinghai. Thousands of people lost their lives. Our Chinese colleagues dropped all their work and dedicated themselves to rescue efforts. When the dzud led to several million livestock dying in Mongolia following the winter of 2009-10, our team dedicated effort and funding to assist our community partners in rebuilding their livelihoods.

These, and other potential situations such as a drought, or a disease outbreak, are unexpected, acute and episodic situations. Helping communities in such emergency situations – irrespective of any biodiversity linkage – is a given, and a humanitarian imperative. As I have discussed throughout, community-based conservation is about relationships. What good is a relationship that cannot come to the assistance of a community in distress?

These are serious situations. What if the issue may be episodic, but not as serious? Sometimes, agreeing to assist the community with issues unrelated to biodiversity could help strengthen the relationship substantially. However, being creative and helping them meet their unrelated needs through interventions that are actually biodiversity-linked may be a much better approach than providing direct assistance for biodiversity-unrelated issues. I will explain this with an unlikely example.

During the establishment of our first village reserve, the community had requested advance payment for two years, which we had agreed to. It turned out that they used the money for cooperatively repairing the village temple, and improving access to it.

Because the funding was used, from their perspective, for a noble cause, it ended up helping our relationship quite a bit. The funds they had received, though, were in lieu of the village reserve, and not specifically for temple repairs. They were, of course, free to use the funds in the way they desired. This experience helped to understand that such unrelated needs can potentially be incorporated in conservation negotiations with the community (see Chapter 6: NEGOTIATION).

Deciding whether or not, or to what extent, to respond to biodiversity-unlinked societal issues is more difficult when the problem is chronic. Multiple issues become important, but a few that stand out and perhaps merit some consideration here include the seriousness and resource needs of the issue, our expertise (or lack of it), our resources, and the risk of creating undue expectation. As we will see, none of these is simple.

Healthcare is a good hypothetical example to consider. It is a critical need, and though unrelated in many ways, it does determine the ability of a community to participate in conservation and most other individual and societal pursuits. It is serious enough to make it worthy of consideration. How then can we assist a community living in an important biodiversity area with a basic need such as healthcare? Should we?

One of the important things to consider here is that healthcare is a sustained, resource-intensive need. So we need to be clear that unless it is an episodic event (such as an epidemic outbreak), an ephemeral involvement is not really of much value for the communities, and may end up creating expectations that we are unable to meet.

Secondly, do we have the expertise? Healthcare is a technical subject. There are also ethical issues involved. For example, having to close down a healthcare program after having created expectations – and without an alternative in place – is perhaps a worse situation to be in than not having been involved in the first place.

Unless we have public health specialists on our team, it may be prudent not to get involved directly. If the issue is important enough, the best way to assist the communities might be by encouraging and assisting specialized healthcare organizations to come in and establish public health programs.

Indeed, it is useful to be aware that even helping to set up a collaboration like this requires a fair amount of time and effort. When the community's needs are chronic, the effort needed is much greater, and decisions must balance empathy with prudence.

Another aspect to keep in mind while making decisions on unrelated interventions is the nature of overall partnership with a community. As the relationship with any community matures, unrelated interventions become easier to manage in terms of expectations, and become more useful in strengthening the relationship.

This is in contrast to the standard practice of community entry activities in social work. Here, unrelated needs are sometimes considered under the broad categorization of community entry activities. I would suggest, instead, that community entry is best served through communication, information gathering, and biodiversity-related interventions, rather than unrelated ones. It can otherwise send incorrect messages to the community, and create false expectation right from the outset.

It is also useful to keep in mind that those unrelated activities, including exposure visits and training, that have a greater ability to encourage collective work, skills enhancement, and general improvement of social capital, are to be preferred over those that don't. As I have mentioned before, the conservation threats and the socio-ecological situations change over time. It is the combination of our community relationships and their social capital that is important in determining the resilience of community based conservation programs.

Monitoring and adaptive improvement

Evidence for the effectiveness of community-based programs in achieving biodiversity conservation remains limited. On what basis do we decide whether or not a conservation project is having the desired impact? Monitoring and adaptive improvement are fundamental parts of any conservation effort.

> Monitoring has a role in helping describe a conservation program comprehensively and quantitatively, to measure its quality and impact, and to assist in improving or even creating suitable conservation interventions.

Monitoring has a role in helping describe a conservation program comprehensively and quantitatively, to measure its quality and impact, and to assist in improving or even creating suitable conservation interventions.

As mentioned earlier, biodiversity goals may not be achieved at times, either because the implementation of conservation interventions was not done well, or because the interventions did not address the main threats, or due to external limiting factors. A good monitoring program helps diagnose where the problem lies, and accordingly, allows for adaptive improvement.

What should we monitor? In community-based conservation efforts, there are typically three types of indicators that we need to monitor to varying degrees.

These include (i) keeping a constant watch on the nature and severity of threats to biodiversity in any area, (ii) process indicators that help assess how well the conservation interventions are being implemented, and (iii) impact indicators that help assess the actual impact of conservation interventions on the biodiversity that one is trying to conserve.

A good monitoring program relies on a variety of indicators, because biological variables such as snow leopard populations, prey populations, and the status of rangeland vegetation, convey different kinds of information compared to say poaching records, or the extent of livelihood generation for community members.

The information content, utility and robustness of different indicators varies. For instance, biological indicators tend to have greater natural variation and longer response times, making them more difficult to interpret on their own. Therefore, it is useful to combine their monitoring with other indicators such as changes in peoples' attitudes, or other measures of threats reduction.

The techniques to quantify and monitor the various indicators vary, as do the periodicity and spatial extent over which each indicator needs to be quantified.

For instance, estimating snow leopard abundance in key conservation landscapes once in 2-3 years may be desirable and feasible, while it may be possible to undertake prey abundance monitoring in representative parts of this landscape once in two years or even annually. Some program indicators such as the number of local livestock insured, or the extent of livelihood generation supported, are best monitored at least on an annual basis.

It makes sense to measure some biological indicators such as the snow leopard population at larger scales, and it is difficult to relate them to a specific community conservation intervention. Indeed, such monitoring is best done in collaboration with the Government – or to at least share the results of such efforts with them constantly. It is, after all, the Government that is the overarching body usually administering larger landscapes with various land tenures such as community land and protected areas and various stakeholders (Chapter 9: STRATEGIC SUPPORT).

If a community intervention is designed to influence the prey population, the status of vegetation etc., it would be sensible to monitor those indicators at the level of community-owned land.

Monitoring can be resource – and manpower – intensive; and therefore, the extent of monitoring is often decided by logistic constraints. Within these constraints, though, and much like community-based interventions, monitoring too is best done in as multi-pronged a manner as possible. It ideally focuses on a combination of biological, socio-economic and attitudinal indicators; some of which may describe program implementation while others help evaluate conservation impact.

Dos:

- Monitoring threats, interventions and impact
- Adapting and improving interventions whenever possible or necessary
- Helping communities when they have urgent, episodic needs unrelated to biodiversity
- Looking for ways to assist communities in biodiversity unrelated needs with interventions that are linked to biodiversity

Don’ts:

- Assuming that threats and priorities remain stable

- Forgetting that problems are opportunities to improve conservation interventions
- Creating expectations that one cannot meet
- Getting directly involved in biodiversity unlinked interventions when the team lacks the necessary expertise

Chapter 9:
STRATEGIC SUPPORT

While local communities are made out to be a major problem for biodiversity conservation, it is really the larger – often global – economic forces that usually overwhelm conservation efforts today. Governments alone have the main authority to bring about a greater balance between the needs of biodiversity conservation and those of economic development. Ultimately, the success of community-based interventions depends, to a large extent, on the support of the government. In the absence of supporting policies, laws, or political will, years of community-based conservation effort can easily be laid to waste in the face of economic forces. Practitioners must work closely with governments in policy formulation, management planning, and implementation, and in catalyzing multi-sectorial cooperation. This role requires a delicate balancing act where the practitioner must cooperate and partner with governments, and at the same time oppose them when warranted in the interest of biodiversity conservation.

Community-based conservation is a demanding undertaking, in terms of time, resources, effort, and perseverance. As we have seen, there are constantly emerging issues that require attention, ranging from new threats to biodiversity or the failure of interventions to have the desired impact, to problems arising out of unrelated local conflicts or politics. For the conservationist trying to engage in community-based conservation, the plate is constantly brimming over.

Yet, it is critical that we are able to think beyond the next incremental decisions and the day-to-day contingencies. This is necessary because although local communities are often made out to be a big problem for conservation, it really is the larger external economic forces that globally overwhelm conservation efforts today.

Retaliatory killing of snow leopards, for instance, is a much easier problem to have and to manage compared to e.g. the expansion of mining into important snow leopard habitat. That's particularly true in a situation where more than 10 % of the GDP of a country comes from the operations of a single mining company. Extreme as it may sound, this is not an imaginary example. It accurately describes a real situation that we have wrestled with.

> Although local communities are often made out to be a big problem for conservation, it really is the larger external economic forces that globally overwhelm conservation efforts today.

Routinely today, conservation goals are pitted against global economic pressures, and they are routinely compromised at the global, national, and local levels. In such a scenario, years of community-based conservation effort can be laid to waste in the face of strong economic forces, or in the pursuit of seemingly legitimate national and local development agendas.

This is where strategic support for community-based conservation becomes so critical. And it comes from an unlikely ally.

Governance and the government

As we have seen throughout this document, community-based conservation is founded on the ideas of equity, devolution, and local empowerment. It aims to shift the responsibility of conservation from solely resting with the government to a governance model where local communities play a central role in conservation.

However, it would be a mistake to view community-based conservation as a zero sum game, where an increased role of local communities translates to any reduction in the role of the government. To the contrary, we desperately need conservation to acquire a higher place in the government order today, more than ever before.

The illegal trade in wildlife products, for instance is already estimated to be over US$ 20 billion annually (Graham-Rowe 2011). The scale and manner in which it needs to be tackled, especially in terms of enforcement, can only be effectively handled if governments and international alliances put in the needed effort. The scale of climate change, similarly, requires governmental leadership, integrative international negotiation, and stricter legislations.

Even at the local and regional levels, the role of governments in biodiversity conservation remains integral. While infrastructure and development projects are necessary for any nation's growth, such projects become problematic when they are located in important biodiversity areas, or are detrimental to the welfare of local people.

> In situations where infrastructural projects are a threat to both people and biodiversity, conservationists and local communities can together form an influential force to resist or ameliorate them, provided they have a conservation partnership based on mutual trust.

In situations where infrastructural projects are a threat to both people and biodiversity, conservationists and local communities can together form an influential force to resist or ameliorate them, provided they have a conservation partnership based on mutual trust. Ultimately, however, the decisions, one way or the other, rest with the government.

We were able to make progress with establishing a Protected Area in Tost Mountains of South Gobi, mentioned earlier, thanks to multiple factors and circumstances: the local community was united in their opposition to mining expansion; Tost represented an important snow leopard habitat so we were determined to help protect it; we had a history of collaboration with the local community through Snow Leopard Enterprises (Chapter 10); there was a bedrock of scientific information to demonstrate the importance of the area; and in a pre-election year, there was unprecedented support and pressure from within Mongolia to protect the area as a tribute to Sumbee, our young colleague working in Tost who passed away in late 2015.

But ultimately, it was the government alone that had to decide whether or not to approve the proposal, first to declare Tost a Local Protected Area, which it did in 2010. Similarly, it was up to the Great Hural, Mongolia's parliament, to decide whether or not to upgrade the Tost Local Protected Area to a State Nature Reserve. It chose to do so in 2016. It could have chosen not to.

In a similar manner, when we joined hands with other conservation and human development NGOs to oppose gold mine expansion into Kyrgyzstan's Sarychat Reserve, our collective effort and our experience in the area came in handy. However, the main reason underlying our success was not the strength of our collective voice, but the favorable political circumstances prevailing at that time in the Kyrgyz Parliament that helped our concerns to be heard. The government chose to consider our concerns on that occasion. At another time, it could have easily disregarded them.

There are things we can do proactively, however, beyond just hoping for favorable political circumstances whenever such problems arise.

Policy and management plans

How can we try to bring more balance between the needs for economic development and biodiversity conservation? How can we create more space and support in government thinking for community involvement in conservation? The answer is relatively simple, but getting there is extraordinarily difficult.

It is essential for us to work closely with governments to create supportive processes and structures within the government system. These need to facilitate more rational decisions that better balance economic development needs with the needs of biodiversity. They also need to strengthen the voice of communities in such decision-making – in reality, and not just in rhetoric.

It requires changes in policy, appropriate management planning and implementation, and, ideally, as we will see later, a stronger legal system in support of community-based conservation (see Chapter 13).

Policy and management planning generally tend to be viewed as lying completely within the purview of the government. Yet, there is the space and the need for conservationists to be centrally involved in policy planning and implementation– and there are numerous examples as well.

India's Project Snow Leopard, a national strategy and action plan meant to guide conservation of high altitude biodiversity in all five Himalayan provinces of the country, was a product of years of our collaborative effort with the Central and Provincial governments. We catalyzed the process, drafted the document on behalf of the government, and lobbied for its official endorsement. Similarly, as we helped catalyze the Global Snow Leopard and Ecosystem Protection Program, aimed at the highest levels of all 12 snow leopard range-country governments, our teams assisted various national governments in creating their related strategies (called National Snow Leopard and Ecosystem Protection Program).

Needless to say, the global program mentioned above as well as the national program in India recognize a central role for local communities in conservation and conflict management programs; and they facilitate collaborations among local communities, wildlife managers, and conservationists. They adopt a landscape-level approach to conservation that looks well beyond the boundaries of Protected Areas.

Similarly, we have assisted some of the Indian snow leopard provinces in the identification of important snow leopard landscapes to be brought under comprehensive, community-based conservation efforts. In some of the provinces, our teams have helped create landscape-level management plans. We are now assisting the wildlife managers in implementation, especially in engaging better with local communities (see Chapter 13), and trying to catalyze multi-sectorial cooperation.

While the government continues to remain a key player, a greater emphasis on community-based conservation does imply some realignment and refocusing of its approach at the local level. In this approach, wildlife managers, rather than relying on their own limited human resources, try to achieve conservation in cooperation with local communities who assume a dominant role in conservation micro-planning and implementation. The interaction is mediated by conservationists.

Our catalytic efforts are helping to bridge the distance between wildlife managers and local communities, a relationship that has traditionally tended to be edgy. Improving it will in turn, we believe, make conservation efforts more resilient, and will improve our collective ability to negotiate when external forces threaten to destroy local biodiversity.

Multi-sectorial cooperation

The distance, however, is not just between local communities and wildlife managers. It also exists between the various departments of the government itself. For landscape species like the snow leopard, as I have discussed earlier, a Protected Area approach is ecologically insufficient (see Chapter 3: APTNESS). But in any land outside of Protected Areas, there are numerous stakeholders, including several administrative bodies. Various government departments have a role here, such as those responsible for building roads and infrastructure, for conservation, agriculture, etc., whose mandates are often conflicting with each other.

Conservation suffers because these departments don't talk to each other as much as they should, and because the mechanisms for inter-sectorial communication within the government tend to be poor. Our colleague Yash Veer Bhatnagar has spent years in India trying to get various government departments to cooperate for conservation. Using his experiences, he now assists other range countries in their management planning and multi-sectorial cooperation.

We have learnt from experience that facilitating better communication and cooperation between various government departments can help not just better safeguard conservation interests, but can actually assist in generating more resources for community-based conservation as well.

> Facilitating better communication and cooperation between various government departments can help not just better safeguard conservation interests, it can actually assist in generating more resources for community-based conservation.

For example, as free-ranging dogs are becoming a serious threat to wildlife and a human health and economic hazard in many snow leopard landscapes, we have been able to initiate a pilot program in cooperation with various government bodies including the wildlife and veterinary departments, the district administration, and the local communities, to address the problem. As a result of this cooperation in the Western Trans-Himalaya, our colleague Ajay Bijoor has been able to help channelize the expertise and resources of various departments – rather than solely taxing conservation funds – for activities like dog sterilization and vaccination, and garbage management.

Thus, through appropriate management planning and actions, conservationists can catalyze collaborative multi-sectorial efforts for biodiversity conservation and human welfare. However, such multi-sectorial cooperation depends on the government's willingness, underscoring again the fundamental role of the State in community-based conservation.

The art of finding middle ground

Conservation is the art of finding meeting ground amidst conflicting interests and priorities. It is about tradeoffs between the need to protect biodiversity and the need for development and prosperity. It is about finding effective solutions through integrative negotiations.

In almost every case, we need to compromise to a certain extent. We can improve the resilience and sustainability of community-based efforts by strategically partnering with the government. By generating strategic support of the government, we improve the chances of tilting the balance in negotiations in favor of biodiversity conservation.

But, as discussed earlier, the nature of the problem is such that no amount of effort or strategic support, can guarantee positive outcomes for biodiversity and human welfare. Indeed, under the pressures of economic development, policies are sometimes ignored, and even laws are circumvented or broken by the very same bodies that are responsible for creating, implementing, or upholding them.

Working with governments can be frustrating, just like it can be occasionally with local communities. It tests the conservationist's patience, perseverance, and negotiation skills. Conservationists are in an unenviable position where they must collaborate with the government and oppose it at the same time when warranted in the interest of biodiversity conservation. Good diplomacy and negotiation skills can help traverse this delicate path. A set of PARTNERS Principles for effectively working with governments is much needed.

For conservation efforts and impact to be sustainable, strategic support of the government is essential. If we want to enable local people to have a strong voice in conservation, paradoxically, we must invest time and effort into working with governments. If we are unable to make this investment, community-based efforts will not get the strategic support they need, and external economic pressures will easily overwhelm conservation efforts and goals.

Returning to the communities themselves, the next three chapters (Chapters 10-12) provide descriptions of three specific community-based initiatives that the Snow Leopard Trust has been involved in. These are written from the perspective of the practitioner who might be considering piloting such an initiative themselves. None of ours is perfect. We try to improve as we go along.

Dos:

- Proactively collaborating with government and sharing expertise
- Facilitating cooperation and communication between various government sectors
- Acting as a bridge between local communities and wildlife managers

- Compromising and reconciling, while being prepared to oppose the government when it is warranted

Don’ts:

- Viewing the government as anathema for community-based conservation
- Assuming there is no role for the practitioner in policy formulation, management planning and implementation

Part II: Description of specific community-based initiatives

Chapter 10:
SETTING UP CONSERVATION-LINKED LIVELIHOOD MECHANISMS: CASE STUDY OF SNOW LEOPARD ENTERPRISES

Contributions: Agvaantseren Bayarjargal, Brad Rutherford, Gina Cantara, Jennifer Snell Rullman (Program implementation, text, and information), Kulbhushansingh Suryawanshi (Information compilation)[1]

Sharing landscapes with carnivores such as snow leopards and wolves imposes costs on local people: there are economic setbacks because predators kill livestock, there is fear, and there are material and opportunity costs of guarding livestock. These costs often generate resentment and lead to conflicts between livestock production and carnivore conservation. Helping increase peoples' tolerance and enhancing their ability to coexist with predators is necessary for large carnivore conservation.

Education and awareness programs as well as economic incentives can be useful in improving peoples' tolerance towards wild species in conflict (Mishra and Suryawanshi, 2015). Incentive programs try to enhance local livelihoods in a way that benefits focal species or habitats (Mishra et al. 2003a). They are different from measures such as insurance or damage compensation that try to directly

[1]*Citation:*
Bayarjargal, A., Cantara, G., Snell Rullman, J, Rutherford, B., Suryawanshi, K.R. and Mishra, C. (2016). Setting up conservation-linked livelihood mechanisms: case study of Snow Leopard Enterprises. Pp. 109 to 126 in: Mishra, C. (Ed.) The Partners Principles for community-based conservation. Snow Leopard Trust, Seattle, USA.

share and offset the economic losses caused by wild species (Figure 10.1; also see Chapter 11). Incentive programs act to improve peoples' tolerance towards wildlife by strengthening livelihoods through measures that are directly or indirectly linked with conservation.

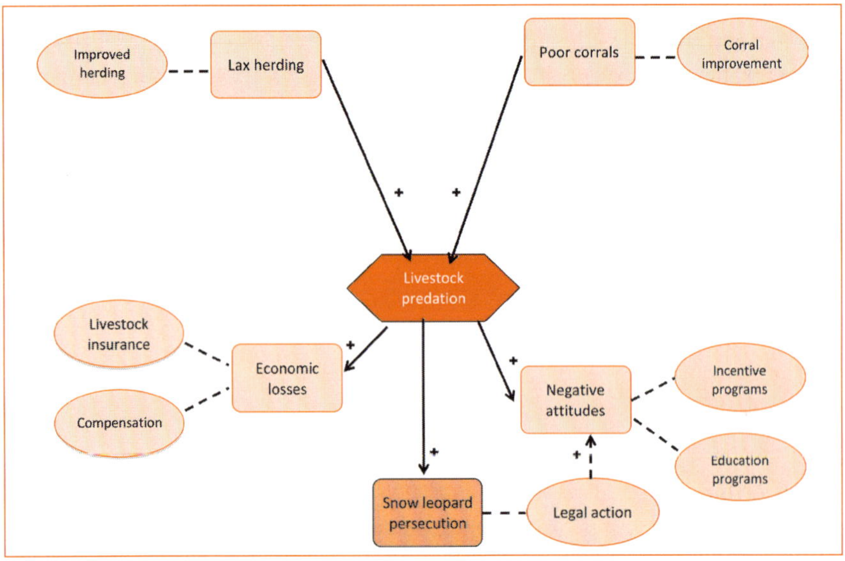

Figure 10.1.: The various anthropogenic causes and effects (rectangles) of livestock depredation by snow leopards and sympatric carnivores, and the role of different kinds of interventions (ovals) in managing conflicts over livestock depredation (modified from Mishra et al. 2016b).

Snow leopards are landscape species and come into contact with people and livestock, often irrespective of whether the community resides inside, on the fringes of, or far away from a Protected Area. Across their range, retaliatory killing in response to livestock depredation is an important cause of the endangerment of snow leopards and wolves. Conservationists must work with local communities and help strengthen their ability to coexist with predators.

Snow Leopard Enterprises (SLE) is a conservation-linked incentive program established initially in Mongolia by the Snow Leopard Trust and its partners. It has subsequently expanded to other countries including Kyrgyzstan, Pakistan, and more recently, India. SLE focuses on production of handicrafts by households living within snow leopard habitat, primarily using locally produced material. Generally, it involves the use of a small proportion of the wool produced by the participants from animals such as sheep, camel, or yak. Products include rugs and other home décor items, ornaments, toys, cat toys, slippers, baby booties, hand bags, eyeglass cases, etc. Over time, other sustainable materials such as jute have been added to the program that was originally primarily focused on wool.

The bulk of the products are marketed and sold by the Snow Leopard Trust's sales and marketing team through conservation partners and distribution channels in North America and Europe.

SLE has proven to be a useful tool for improving peoples' tolerance towards snow leopards and for garnering their support for conservation by helping build partnerships. SLE is also a potentially useful biodiversity-linked community entry program (see Chapter 7: RESPONSIVENESS), as starting SLE in a new community is less contingent upon pre-existing trust and social capital (see section below: SLE and the PARTNERS Principles) compared to other initiatives, such as our livestock insurance program (Chapter 11).

SLE provides local communities with opportunities for livelihood enhancement and diversification in exchange for a commitment towards wildlife conservation. While a program like this could help against a variety of threats to biodiversity, as we shall see, our current SLE program specifically aims at a complete curtailment of hunting of snow leopards and their prey on the respective community's land. The community takes responsibility to prevent hunting by its members and by outsiders. A built-in bonus system qualifies the participants to claim an additional 20% on the handicrafts income if the conservation commitment is honored during the year.

In Mongolia, in about half the communities so far, the participants have agreed to provide 3% of the bonus amount into the general fund that each herder community tries to maintain for community work. This creates a strong incentive to prevent hunting, as any violation of the conservation contract leads to a complete loss of bonus on that year's handicrafts income for the entire community.

The community takes responsibility to prevent hunting by its members and by outsiders. A built-in bonus system qualifies the participants to claim an additional 20% on the handicrafts income if the conservation commitment is honored during the year.

We invest an amount equivalent to 10% of the total handicrafts purchased from each province in Mongolia into a provincial-level fund from which small grants are awarded for conservation-friendly activities. These conservation grants are open to community members, schools, protected area staff or environmental agencies etc. They encourage and support innovative conservation action. The grants program was started in 2007, and between 2008-2014, 50 of the approximately 90 proposals were awarded. These have included a community's effort to clean up a local stream, the production of a field guide of the wildlife of a protected area, various conservation awareness initiatives such as eco clubs, and wildlife surveys by protected area staff.

Program in operation

SLE was started in two snow leopard regions of Mongolia in 1998 with 50 families, belonging to 8 communities. It is currently operational in seven provinces along the Altai Mountains with 225-250 participating families in 31 communities. Since 2000, more than 700 individual community members have benefitted from SLE (Figure 10.2).

> The average contribution of SLE to each participant's income was around US$ 55 in 2005 and had increased to approximately US$ 150 by 2013.

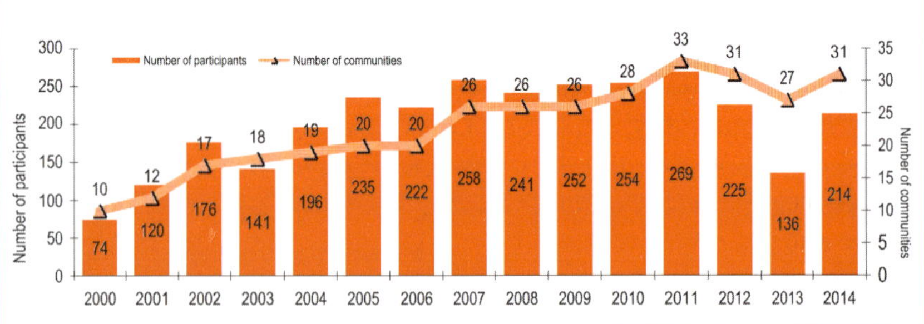

Figure 10.2. Trends in the number of communities and participating households in Snow Leopard Enterprises in Mongolia.

On average, 64% of the families in each community currently participate in SLE. The average contribution of SLE to each participant's income was around US$ 55 in 2005 and increased to approximately US$ 150 by 2013. While the average per capita income for low and middle income families went up in Mongolia from US$887 in 2005 to US$ 2073 in 2013, SLE income has continued to contribute a significant proportion of cash income to the far-flung rural communities that we partner with.

Between 2000 and 2005, the number of SLE communities increased from 8 to 24. During this period, 12 instances of hunting were recorded: 3 of snow leopards and 9 of wild ungulates. In all cases, the participants and communities had to forgo their annual bonus. Between 2006 and 2014, the number of communities increased to 31 (Figure10.2). An additional 6 communities are now in their first years of piloting SLE. From 2006 to 2014, 5 instances of hunting were recorded: 4 of snow leopards and one of a wild ungulate.

Whenever the bonus has been lost, the community has taken action, creating high pressure against hunting. In 2008, for instance, a herder who killed a snow leopard in defense of his livestock had to pay the participants the bonus equivalent of US$860 that they lost because of his action. In another case, where a red deer was killed, the SLE participants found out who the hunter was, demanded compensation for the lost bonus money from him, and invited his wife to join SLE.

Whenever the bonus has been lost, the community has taken action, creating high pressure against hunting. In 2008, for instance, a herder who killed a snow leopard in defense of his livestock had to pay the participants the bonus equivalent of US$860 that they lost because of his action.

In 2014, the total SLE sales managed by the Snow Leopard Trust exceeded $110,000, a 16% increase over 2013. Sales have been close to or exceeded $100,000 per year in 6 of the last 7 years, and have totaled over $1 million at the end of 2014. (Figure 10.3) As mentioned earlier, SLE has helped participants increase their income by an average of $150 per family and has allowed us to partner with communities to improve the protection status in 17% of Mongolia's snow leopard habitat.

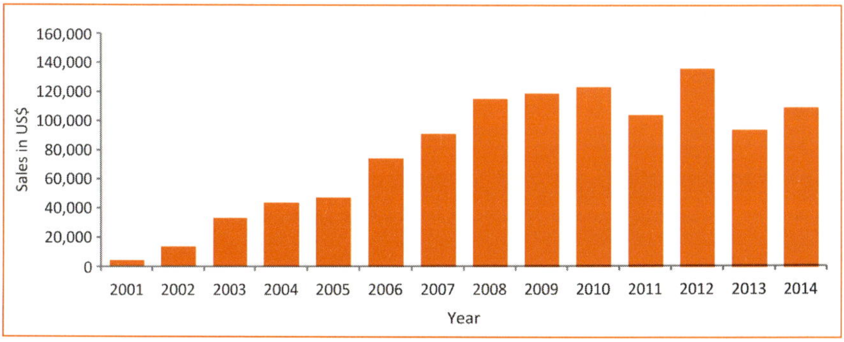

Figure 10.3. Sales figures for Snow Leopard Enterprises since 2001.

In our experience, there can be conflicting interests and needs between the program's conservation aspect and the realities of the market. E.g., certain products may be important to have in the portfolio because they are relatively easy for newer participants to make, even though demand for them may not be as great as the supply. For other products, market demand may be greater than the number of products that participants can supply. In such cases, it is critically important that the conservation aspects of the program are accorded priority.

There is room to grow sales if quality products are available from the field. Approximately 50% of 2014 sales came from selling SLE products through our website and at events. The other 50% came from selling to retail outlets such as pet stores and zoo stores. SLE program revenue is currently sufficient to cover all direct costs such as product purchases and conservation bonuses, as well as shipping and marketing costs, but it can't cover staff time.

Due to its success in Mongolia, the program has been expanded to snow leopard habitats of Kyrgyzstan (46 participating families in 3 communities by 2015), Pakistan (69 families in 2 communities), and more recently, India (46 families in 2 communities). Although the model is similar in each country, it has variations in governance, product portfolio, SLE community organization structure, and some other details depending on the needs, resources, and social and cultural setup.

What it involves: steps in setting up Snow Leopard Enterprises

*Biodiversity surveys and socio-economic assessments are conducted. Surveys assess the conservation importance of the area and help understand the threats to biodiversity, livestock rearing and pasture use practices, and wool production practices such as current levels of processing, skills, equipment, income sources, etc.

*One-on-one discussions are conducted with a few knowledgeable people and local champions to share program details, assess local issues, and gauge their interest.

*Community meetings are held to explain the program specifics and receive people's inputs, and initial discussions take place to assess interest and feasibility.

*Several meetings and informal interactions usually take place before a pilot begins. When a community is ready to join SLE, formal meetings are conducted to discuss further program details. Participants are requested to organize themselves and designate a committee of one or more local coordinators who facilitate the program and serve as the main points of communication.

*During the following year, skill sets are assessed, initial product development takes place, and wool processing and quality control trainings are imparted. These training workshops also facilitate information exchange and discussions on program details.

* During this initial period, a few products are generally ordered and purchased to encourage the participants. Conservation connections and interests are reiterated, but conservation contracts are generally not used yet. The community and the conservationists use this period to evaluate if the program can be run effectively. The level of community coherence, extent of their willingness to commit to conservation and continuity of product production during the pilot phase are considered. If the community doesn't appear to be ready, more time and effort are invested into communication and trust building.

*There is close collaboration between the Snow Leopard Trust sales and marketing team and the field teams to determine marketable product designs. Market surveys are conducted nationally and internationally to assess product feasibility and the product portfolio is modified accordingly. The portfolio evolves over time, based on market demands and relevant training.

*While determining the products, accessibility and environmental sustainability of raw materials are also considered. For example, a product idea to develop chopsticks was dismissed, as we couldn't ensure the sustainable sourcing of wood. Similarly, we have avoided creating buttons from goat horns because it would be difficult to separate them from wild ungulate horns.

*Materials of local origin are preferred. If additional materials need to be purchased elsewhere, the participants either procure them on their own, or our local teams supply them. Participants are given the option of paying for the material cost in kind by providing an equivalent of SLE products instead of cash.

*Pricing is key if SLE is to serve as a conservation incentive for communities, and also for subsequent sale of products. For every product, it is best to assess a reasonable sales price in selected markets, including local ones. To assess the price that can be paid to the producers, we work backwards through the sales, marketing, shipping, and production costs, and include fair labor costs. This is used as a benchmark for negotiations, and if a mutually acceptable price can be agreed upon, the product is included in the portfolio. If there is no agreement, we consider other products.

*Once a product has been determined and price negotiated, further training takes place for quality processing and specific product development. It is useful to develop a guidebook that includes a step-by-step description of how to make the product, tips on what to avoid, clear statements describing benchmark quality for purchase, and photographs of the quality product.

*It is important to conduct periodic skills sharing events and trainings. When new communities join, it is advisable to facilitate an exchange with members and local champions from existing SLE communities who can share their experiences about the program.

*Once the participants have learnt to make quality products, and the community is ready to partner in the conservation program, a conservation contract is signed by the participants, the community, the conservationists, and, if relevant, the Environmental Inspectors or Protected Area staff (Appendix 10.1). A product order is placed, and product purchase time and amounts are agreed upon in the contract. The responsibilities of each party are listed. The contract clearly states the conditions for awarding the compliance bonus. It also explicitly clarifies that in case there is any poaching or violation of the contract, the entire community loses the bonus amount.

*The area over which the community would be responsible for conservation is mapped. If the area of responsibility already exists, it is helpful to confirm the boundaries with the community and distribute maps to all individuals. If it doesn't exist, conservationists help the community determine areas of responsibility by facilitating a mapping exercise for pastures, sacred sites, water sources, vegetation, important wildlife areas, etc. It can take several meetings over a long time period to get community ownership for boundaries of the community-responsible area.

*The coordinators and/or committee are asked to ensure equitable distribution of the order among all the participant families. For transparency, is helpful to share a copy of the total order with all participants.

*In addition to the community conservation contract, an individual SLE participant contract for each household is signed. It specifically states the agreed upon order (e.g. 20 rugs, 15 cat toy mice, 30 chair mats), and the purchase price for the order. It also contains a calculation showing the bonus amount to be received on the order should the community fulfill its annual conservation commitments.

*Monitoring for compliance of contracts and collection of data on livestock depredation and poaching are done with the involvement of community members, key informants, protected area staff and rangers. Where possible, we also encourage third party surveys and research team surveys. The conservation bonus is paid at the end of the working year based on compliance.

*Handicrafts are purchased annually or half yearly at mutually agreed-upon times.

*The conservation contract is renewed and signed during the purchase trip at the end of the current working year. Where necessary, the contract is amended to meet changing threats or program details. For example, in the Kyrgyz Tien Shan, conservation contracts were recently amended to reflect the SLE communities' agreement to not only refrain from hunting themselves, but to not provide housing or aid to illegal hunters coming to the area from elsewhere.

*Purchase records are maintained, which include the number of participants, the number and types of products ordered, the number and types of products fulfilled and purchased, total income from purchase, etc.

*The bulk of our products are shipped to the SLT US office, from where they are marketed and sold by the sales team. Products are primarily marketed online and sold via SLT's webstore; but are also distributed through wholesale partners such as zoo stores and niche markets, and through events such as trade shows, relevant conferences and festivals or holiday sales outlets.

Specific good practices that help avoid problems:

*SLE by itself is not adequate to address the threat of hunting of wildlife including snow leopards and prey. It is best combined with other initiatives as part of a multi-pronged conservation strategy (see Chapter 3: APTNESS).

*It is essential to maintain a strong awareness and outreach component as part of the program, so that SLE is viewed as a conservation initiative, and not solely a livelihood enhancement project. It is important to continue to reiterate the conservation goals of the program to the participants. Otherwise, despite the system of conservation contracts and bonuses, the program is at risk of being seen as an income generation project and its conservation aspects tend to fade in people's understanding.

*It is important to sign contracts at community meetings so that all participants understand conditions of the contract, are aware of the number of products being ordered, and the distribution of orders can be equitable (see Chapter 5: TRANSPARENCY).

*Individual purchase contracts should reiterate the full responsibilities of both parties, and should state the amount of money the individual would receive for the products. It should also state the amount to be received as bonus, provided the conservation commitment is fulfilled.

*Communication must not be restricted to the coordinators. Conservationists must remain in touch with participants as well as non-participants and other community members to the extent possible and obtain their periodic feedback on the program (see Chapter 2: PRESENCE).

*The local coordinators must be chosen by the participants or the community, rather than by the conservationists. Coordination work should ideally be voluntary and not be a paid position, though actual costs should be covered. Helping establish local committees that oversee the local coordinator and the program may be helpful.

*Conservationists must work with coordinators to ensure a transparent and equitable distribution of orders among the participants.

*Sometimes, the producer group can be resistant to expansion for fear of losing part of their orders to new participants. A system of rewarding the community fund for recruiting new members can be helpful.

*As mentioned earlier, it is important to jointly map the area over which the concerned community would be responsible for implementing conservation measures. Moreover, it is helpful for each household to have a copy of the boundaries and local landmarks within the monitoring area.

*It is useful to distribute conservation bonuses during community events. In Mongolia, we typically do this during the Tsagaan Sar (New Year) community celebrations.

*Production of high quality handicraft is a long-term process for non-artisans. It is important to provide skills improvement and capacity building trainings on a regular basis. Providing feedback on product quality after each purchase is important for improvement.

*While placing orders, leaving design samples and instruction guides with the participants is important for quality control. Quality is assessed based on look, design, softness, flexibility, durability, color, etc. Quality control is essential so that the products can compete in the open market. Rejecting low quality products helps to improve quality of handicrafts. At the same time, some sub-standard products made by new participants may be purchased initially as it helps build trust, enthusiasm and confidence. However, it is important to provide feedback and training for improvement of handicrafts, and to clarify that purchase of low quality products would not be possible in the future, as they are not marketable.

*Microcredit loans may be offered to the community to purchase needed equipment for handicraft making. Repayment can be in cash or kind (by handicrafts), and agreements are made for repayment, typically within three years. Gifting of equipment is best avoided, though cost-sharing may be considered.

*Assessing the capacity of the participants through discussions about the number of different products they can produce within a given time is important, so that they can budget their time and capacity. Participants often tend to overestimate the number of handicrafts they can actually produce.

*When adding new products, it is important to consider the price of currently made products. If the new product has a higher price, it may result in a loss of interest in the production of some of the current products. Cost and margins associated with each product also need to be considered for equitable distribution of orders. It is important to diversify product portfolio by skill requirements, time required to make the products, and product price and margins.

PARTNERS Principles and Snow Leopard Enterprises

> Because SLE builds on local, traditional skills and available resources, the concept is familiar and helping communities understand the idea of SLE is relatively uncomplicated

Participating people and communities derive economic benefits from SLE in a fairly uncomplicated manner, making the program relatively simple to implement, though it takes considerable effort. From the perspective of the PARTNERS Principles, while some of the principles, such as Respect (Chapter 4), Empathy (Chapter 7) and Responsiveness (Chapter 8) hold universally and are essential for SLE implementation, some of the others are relatively less essential at the outset. In fact, SLE can actually help strengthen the relationship with the community as it helps address several of the PARTNERS Principles.

PRESENCE

In most communities living in snow leopard habitats, people not only produce wool but have also traditionally used wool to make products for local use, and to a smaller extent, for sale. Because SLE builds on local, traditional skills and available resources, the concept is familiar and helping communities understand the idea of SLE is relatively uncomplicated. Even at the outset, the participants are able to see potential benefits of the program, especially in opportunities for skills enhancement. The investment of the participants is mainly in terms of their time and effort, while the costs are largely borne by the conservation agency. The duration of immersion in the community for relationship building required to initiate SLE is therefore relatively shorter, though long-term immersion is necessary for overall conservation effectiveness (see Chapter 2: PRESENCE) and for developing robust systems of assessing conservation compliance. SLE can be viewed as a valuable long-term engagement tool as well as a useful conservation-linked community entry activity (see Chapter 8: RESPONSIVENESS) that can help build and strengthen the relationship between the community and the conservation agency.

APTNESS

The threats

Our SLE program is mainly designed to address the issue of hunting of snow leopards and their prey. But that doesn't imply that SLE does not have relevance in areas where hunting is not an issue. As mentioned earlier, it is a valuable community entry activity and relationship building tool, and the conservation compliance system can be adapted to address other threats.

The scale

For SLE to have a significant conservation impact, it is important that a substantial section of the community participates in it. That would allow the conservation commitments to be better honored, and a larger resultant contribution to the community fund would also create stronger peer pressure for non-participants to comply as well. In Mongolia, the Government recognizes a group of 15 spatially related herding families as a community, and for this reason, we try to work with communities of at least a similar size. However, SLE can function well even in smaller communities. While SLE must ideally supply a threshold number of products each year based on the size of the market, on the producers' side, SLE is relatively scale independent, and can be started even by very small communities. It is important though that any SLE effort has a built-in system (such as a community fund) that reaches out beyond the participants to the larger community and is not focused solely on the participant families. It is equally important to ensure that SLE and other tools in the conservation portfolio ensure coverage not only within the individual community but also of communities over most of the landscape of interest.

> For SLE to have a significant conservation impact, it is important that a substantial section of the community participates in it. That allows for the conservation commitments to be better honored, and the larger resultant contribution to the community fund helps create stronger peer pressure for the non-participants to also comply.

Values

Conservationists often end up working largely with the men in any community-based effort, as it is often the norm in local communities for men to be involved (or seen to be involved) more in making decisions (see Chapter 3: APTNESS). This is problematic as it leaves out an important, influential half of the community from conservation efforts. We have found that women, presumably suffering a larger burden of the cost of living with large carnivores, tend to have greater negative attitudes towards them (Suryawanshi et al. 2014). While SLE is not gender exclusive, it is one of the rare programs where women members of the family participate much more (98% in Mongolia, 100% in India), bringing them integrally into the conservation dialogue. SLE empowers women economically and socially. With its relatively easier acceptance to communities, potential relevance against a variety of threats, its role in creating a constituency for conservation among women, and, as we shall see shortly, its lower reliance on initial social capital, SLE is a useful – albeit by itself insufficient – conservation tool for most communities under most situations.

> Skills enhancement opportunities in relatively remote snow leopard habitat are still limited, and, not surprisingly, SLE is highly valued for it.

Socio-economics and social capital

As a tool designed to enhance livelihoods while promoting conservation, SLE is likely to be more impactful in less affluent communities. This, however, does not mean that it would not interest communities that are more affluent, because livelihood enhancement, albeit central, is only one of the many aspects of SLE that are appealing to potential participants. Skills enhancement opportunities in relatively remote snow leopard habitat are still limited, and, not surprisingly, SLE is highly valued for it. To start an SLE initiative is easier and not reliant on as much trust and social capital as an initiative such as livestock insurance (Chapter 11). On the other hand, through livelihood addition, skills enhancement, bringing participants together, and creating conservation awareness and peer pressure, an initiative like SLE helps create a fair amount of social capital and instills pride among the participants. That SLE participants contribute 3 % of their bonus amount to community funds in many Mongolian communities is a good example of the outcome of this social capital.

SLE has helped empower local communities in other ways as well. As mentioned earlier, joint mapping exercises are conducted to identify the area over which each community is responsible for implementing its conservation contract. Many communities in Mongolia have used these maps developed under their SLE initiative to negotiate formal land rights with the government, and to claim usufruct ownership over these lands for grazing and other natural resource use, while resisting threats such as mining.

Multi-faceted approach

SLE has shown to be an excellent community entry initiative, and helps improve people's willingness and ability to coexist with predators as well as partner with conservationists. Its generic nature means that it can help against a wide nature of threats to biodiversity. As we have seen in Mongolia, even though it wasn't designed to help with the threat of mining in snow leopard habitats, it has helped communities gain formal recognition of their grazing rights and resist mining expansion.

At the same time, it is important to reiterate that by itself, SLE will not be able to address most of the threats comprehensively, including that of retaliatory killing of snow leopards. As mentioned earlier, that would also require direct efforts to offset economic cost of livestock depredation, such as community-based livestock insurance (Chapter 11), and efforts to better protect livestock such as improved herding and corral improvements. Similarly, it is essential to combine SLE with sustained conservation awareness programs.

TRANSPARENCY

There are some important aspects of Transparency that need to be considered in SLE. First, it's important to constantly clarify and reiterate our conservation goals and the conservation goals of SLE to avoid the program being viewed predominantly as a livelihood generation project.

Next, the purchase price of the various products must be decided in a transparent manner (see below). Further, it is essential to maintain open communication and proactive contact with all the participants and non-participants in a partner community, not just the local coordinators.

Any perception of an unfair or inequitable distribution of SLE orders tends to create discord within the community and must be avoided. It is therefore useful to share a copy of the orders with all the participants and to ensure that orders are distributed equitably.

The role of local coordinators in SLE is both an important and a sensitive one. We have faced situations where local coordinators have come to be viewed as our 'employees', or even as power centers, rather than as fellow community members facilitating the initiative. It is therefore important to encourage the community to select local coordinators themselves, and build in systems (e.g. change or reappointment after a specified term) where local coordinators can be changed if needed without causing discord.

NEGOTIATION

It is understandable if SLE participants wish for higher payments for their products, and one should expect such negotiations to take place. As mentioned earlier, it is best to negotiate the pricing of each product transparently, based on raw material, skill and labor inputs, as well as on the current market value, and transportation and shipping costs – all of which together provide fair standards for the negotiations.

If the participants expect unreasonably high payment, it is useful to reiterate during the negotiations that SLE, in addition to a commitment to fair pricing, provides access to an assured, relatively risk-free market and constant skills improvement – both of which can be very important. At a regional level, maintaining consistency of purchase prices is important, and variation in purchase prices between communities in the same area for similar products should be avoided.

RESPONSIVENESS

When we started SLE, we began by contributing an extra 10% of the value of all products purchased to the nearest relevant Protected Area in the region. However, we realized over time that these funds were not being used efficiently, and in 2007 converted this system into the grants program.

In Kyrgyzstan, discussions with SLE communities have shown that there were a large number of hunters coming to their areas from outside. To reflect this threat to biodiversity, SLE communities have recently agreed to add the clause of not hosting or providing housing to anyone involved in hunting. Such monitoring and course correction are important for any community-based initiative.

Appendix 10.1

General framework for conservation contracts for Snow Leopard Enterprises.

Community name and location:

____

Duration of contract:

____

Name of local coordinator(s):

____

Total number of participants:

____

Order Date: _________________ Purchase Date: _________________

Conservation commitments of participants and community, including clauses on preventing hunting in community responsible areas

____

____

Responsibilities of conservation partner:

*Raising awareness among local people about wildlife conservation

*Monitoring status of wildlife and threats to biodiversity, extent of livestock depredation to carnivores, collating information on poaching etc.

*Purchase of agreed - upon products that meet quality specifications

*Providing bonus to the individual producers, in addition to the purchase value of the products, if no conditions in the contract have been violated

*In case the conservation contract is violated, the conservation partner will be unable to pay bonuses for that contract to any of the participants

*Periodic training in wool processing and manufacturing quality handicraft

*Providing needed equipment through micro credit agreements

Responsibilities of community partner and participants

*Protecting snow leopards and other wildlife from poaching

*Producing products that meet quality and size specifications

*Organizing with other producers in area for self-training and sharing of skills
*Providing finished products to the local coordinator by agreed date
*Documenting and reporting snow leopard and other wildlife sightings, illegal hunting and conservation activities

Product Name	Unit Price	Number	Total

Chapter 11:
SETTING UP MECHANISMS TO OFFSET WILDLIFE-CAUSED ECONOMIC DAMAGE: COMMUNITY-BASED LIVESTOCK INSURANCE

Contributions: Brad Rutherford, Agvaantseren Bayarjargal (Comments and suggestions); Tanzin Thinley, Karma Sonam, Kulbhushansingh R. Suryawanshi, Ajay Bijoor, Radhika Timbadia (Program implementation and updated program information)[2]

When humans share habitats with wild species – especially with large-bodied, potentially dangerous ones – there is usually some detriment to human interests. These come in the form of damage to property, injuries or even loss of human life, and, quite often, loss of livestock due to predation by carnivores (Mishra et al. 2016b).

Offsetting economic costs to farmers due to livestock killing by large carnivores or to crop damage by wild herbivores is a constant challenge for conservationists. Two tools often considered by conservationists are damage compensation and livestock insurance programs. Amongst these, compensation programs have been widely employed, and merit some discussion.

---

[2]*Citation: Mishra, C., Rutherford, B., Thinley, T., Sonam, K., Bijoor, A., Timbadia, R., Bayarjargal, A., and Suryawanshi, K.R. (2016). Setting up mechanisms to offset wildlife-caused economic damage: community-based livestock insurance. Pp. 127 to 148 in Mishra, C. (Ed.) The Partners Principles for community-based conservation. Snow Leopard Trust, Seattle, USA.*

Compensation programs

Damage compensation was one of the early interventions designed to offset economic losses faced by local people due to livestock depredation, and continues to be practiced in many countries, especially in Europe and in parts of Asia such as India (Morrison et al. 2009). It is also prevalent in North America, but since the Federal Government in the United States has avoided getting involved in damage compensation programs, they have been run by NGOs and State Governments. A few programs have been tried out in Africa, though many of them appear to have neither lasted nor met the expectations of farmers (Morrison et al. 2009).

> By attempting to shift the economic burden of conservation from local communities to the society at large, compensation programs, in principle, represent an appropriate means of conservation & conflict management.

By attempting to shift the economic burden of conservation from local communities to the society at large, compensation programs, in principle, represent an appropriate means of conservation and conflict management (Mishra and Suryawanshi 2014). However, unless managed carefully, many compensation programs tend to result in aggravating conflicts rather than mitigating them (Nyhus et al. 2003).

State-run compensation programs often fail to address conflicts effectively due to several factors such as low compensation rates, false claims or corruption, depletion of funds, bureaucratic apathy, and the time and effort required in securing compensation (Mishra 1997; Nyhus et al. 2003; Madhusudan 2003; Maclennan et al. 2009). It is useful to examine the reasons for these failures.

PARTNERS Principles and compensation programs

Indeed, most compensation programs appear to fall short of fulfilling several of the PARTNERS Principles. They are usually centrally managed by a State Department or an NGO, and don't require immersion of the agency representatives into the community (Chapter 2: PRESENCE). The onus is usually entirely on the farmer, rather than the compensating agency, to make contact with the authority and claim compensation. This can create practical barriers especially for those farmers who may be less educated or uncomfortable dealing with officials (Muruthi 2005, Lamarque et al. 2008).

The partnership between the compensating agency and the farmer is usually not equal, and often represents that of a donor and a recipient (see Chapter 4: RESPECT). The programs tend to follow a one-size-fits-all approach, which is often inappropriate in the local context (Chapter 3: APTNESS). The amount of compensation provided is usually very low compared to the value of livestock lost (estimated at 3% in one study; Mishra 1997), and further costs are added for the farmer due to time and effort they have to spend on trying to secure compensation, but also to corruption. (Mishra 1997).

Conservation programs often fail to adapt as the costs of compensation mount with the recovery of the carnivore species (Treves et al. 2009) or due to inflation or changes in market value of livestock.

Compensation programs often fail to adapt as the costs of compensation mount with the recovery of the carnivore species (Treves et al. 2009) or due to inflation or changes in market value of livestock. However, rather than responding to changing realities (Chapter 8: RESPONSIVENESS), compensation programs, being dependent entirely on external funding, have often been stopped as funds have dried up, making the entire effort more damaging than helpful.

When things go wrong in a compensation program – for instance when a legitimate claim is denied or an illegitimate one is honored, or when there is a temporary funds shortage – the lack of ownership among the affected people can cause resentment rather than enabling resolution and course correction.

Furthermore, a certain amount of livestock loss to wild carnivores has perhaps been traditionally acceptable to many local communities. Poorly managed compensation programs tend to reinforce the notion that wildlife is the state's property and that therefore, the responsibility of management of livestock depredation losses should rest entirely with the state (Madhusudan 2003). This also tends to weaken any ownership that communities could feel over such programs, as they are entirely external and not based on discussion and negotiation.

There are other challenges that compensation programs face. They usually operate at large scales, which, together with a lack of ownership, can increase the moral hazard or the temptation to file false claims. Thus, although representing a widely employed conflict mitigation measure, compensation programs by themselves are often unable to change peoples' attitudes towards carnivores

(Naughton-Treves et al. 2003; Gusset et al. 2009), and yet, discontinuing them causes retaliation and hostility (Bangs et al. 1998).

Community-based livestock insurance

Community-based livestock insurance programs have historically been initiated as an alternative to direct compensation (Mishra 1997, Morrison et al. 2009). They involve devolution of authority and assign greater responsibility to the affected communities, and require that in addition to the conservation agencies, the farmer participants also make financial contribution to the insurance fund. In our programs, the need for external funds reduces over time.

Community-based insurance programs appear to work better than damage compensation programs for a variety of reasons, though, as we shall see, it is best to view compensation and insurance as being along a continuum (Mishra and Suryawanshi 2014).

> In our programs, the need for external funds reduces over time. The programs tend to become financially self-sustaining in about 5-7 years.

In our programs described in the rest of this chapter, the need for external funds reduces over time. The programs tend to become financially self-sustaining within about 5-7 years, except during occasional years of excessive livestock losses, when further external financial contributions may be required, or periodically when premium and compensation amounts are to be revised to keep up with inflation or increase in market value of livestock (see Program in operation below).

In snow leopard habitats across Asia, pastoralism is the predominant land use (Mishra et al. 2003a). Snow leopards and other carnivores such as wolves frequently come into contact with livestock. Predation on livestock by these two species results in extensive conflicts between livestock rearing and conservation goals (Mishra et al. 2016b).

The extent of livestock predation by these carnivores is reported to be substantial in many areas, with reported losses averaging between 1.9 to 5 livestock heads per family annually; equivalent to 2.9 to 12 % of local livestock holdings (Oli et al. 1994; Mishra 1997; Jackson and Wangchuk 2001; Namgail et al. 2007; Mishra and Suryawanshi 2014). This translates to significant financial losses for local communities, and, in retaliation, snow leopards and wolves are persecuted throughout their range. The existence of retribution or preventive killing also makes the snow leopard vulnerable to the demands of the illegal trade in carnivore fur and bones, as local killing can quickly form linkages with the more organized illegal trade networks (Mishra et al. 2003a; Mishra and Fitzherbert 2004).

People's attitudes and presumably their behavior towards snow leopards are influenced by a range of factors at various scales ranging from the individual to the community, and include gender, education, income, village size, and livestock holdings and composition (Suryawanshi et al. 2014). Because conflicts over livestock depredation typically have multiple dimensions, multi-faceted conservation programs are required to address them (Mishra and Suryawanshi 2014, Mishra et al. 2016b; also see Chapter 3: APTNESS). In most cases, they require helping livestock owners offset economic losses when livestock depredation takes place; efforts to reduce livestock damage through better anti-predatory livestock management; and improving the social carrying capacity for the predators through conservation-linked livelihood and awareness programs (Mishra et al. 2016b). At the same time, from the conservation perspective, these must be linked to and directly address the threats that the species faces in any area, such as retaliatory killing or poaching.

Program in operation

(Modified and updated from Mishra and Suryawanshi 2014)

We started our community-based livestock insurance program in 2002 in Kibber, a village in India's snow leopard habitat of Spiti. Before launching the program, one of us (CM) had spent several years building relationship and through immersion in the community (Chapter 2: PRESENCE).

The regulations and guidelines for the insurance program were drafted jointly with the livestock-owning families in the village. A committee comprising of four community members was set up. These insurance program committees, whose members typically rotate on an annual basis (or slightly longer in some villages), are responsible for collecting premiums, managing accounts, and maintaining an insurance register. Committees also verify the cause of livestock death.

Decisions on the composition of the insurance committee, the premium amounts (decided based on people's ability and willingness), compensation amounts (aimed at offsetting 50-100 % of the value; see below, Table 11.1), and most other regulations are made and revised collectively. There are clauses that safeguard the interests of conservation by forbidding wildlife persecution, including the collection of carcasses or meat from livestock kills in the pastures.

Per the communities' wish, the program in Kibber and other villages in Spiti and Ladakh was restricted to large-bodied livestock species, and did not include sheep and goat due to their relatively low value locally. When we expanded the program to Mongolia, however, goat and sheep were included in the insurance program (see below).

In Kibber, surveys had shown that the majority of carnivore attacks on livestock occurred in the open pastures, whereas the animals were usually penned inside houses or in relatively secure corrals in most villages (Mishra 1997). The local community accepted that negligent herding was an important cause of these livestock losses, and considered various options to improve herding practice and increase herders' responsibility. As one measure, small monetary rewards (US$ 25 to 40) were instituted for herders who had adopted and implemented good anti-predatory herding practices over a six-month term.

Local knowledge suggested that most of the predation losses of free-ranging yaks, which are usually left to roam in relatively distant pastures, took place when individuals got separated from the main herd. To address this problem, it was agreed that the insurance fund would support stipends for two villagers to herd the straggling animals during the period following yak birth season.

Table 11.1 Premium rates and compensation amounts in the community-based livestock insurance program Kibber, Spiti Valley, India.

Species	Age-sex group	Premium (US$)	Compensation (US$)
Yak	Adult male	0.50	217
	Adult female	0.33	117
	Sub-adult male	0.25	58
	Sub-adult female	0.17	50
	Young male	0.83	50
	Young female	0.67	42
Horse	Adult male	0.50	217
	Adult female	0.50	217

	Young male	0.67	67
	Young female	0.67	58
Cow and cow-yak hybrid	Cow (all ages)	0.25	117
	Hybrid (all ages)	0.25	117
Donkey	Adult male	0.25	42
	Adult female	0.17	25

Of a total of 68 families in Kibber who owned the types of livestock covered under the insurance program, 50 joined the program initially. However, within the first two months, 9 families withdrew their membership, either because they sold their livestock or because they became unsure about the future of the insurance program. The proportion of families participating peaked in 2005-06 when 88% of households participated. Excluding the data from families that withdrew their membership, a total of 172 large-bodied livestock from Kibber were insured in the first year. The number of livestock insured kept increasing till 2007-08, when 273 livestock heads were insured. It has since remained steady in this village (see below, Table 11.2).

On average, participating families from Kibber contributed US$ 16 as annual premiums in the first year. This amount increased to a maximum of US$ 32 by 2006-07 and has since decreased again to US$ 17. Approximately 60% of the total money in the insurance fund collected over the first five years (2002-07) came from outside conservation funds, while the local community contributed the remaining 40 % as premiums. Between 2009 and 2014, however, only 11% of the insurance fund's total volume were contributed by conservation funds in the form of one-time additional support of US$ 420 in 2013-14 (see below, Table 11.2).

Table 11.2: Details of the community-based livestock insurance program in Kibber village from 2002 to 2014

Year	Number of households in the village	No of households participating in the program	Total no of large-bodied livestock insured	Total annual premium collected (US$)	Average premium contributed (US$) by participants	Total number of animals lost	Number of participants receiving compensation	Total compensation paid (US$)	Bank balance of the insurance program (US$)	External contribution to insurance fund (US$)	Average amount received per participant (US$)	Exchange rate: INR per US$
2002-03	72	50	173	912	18	9	8	775	1087	1000	97	40
2003-04	72	41	185	818	20	14	10	1000	1731	1000	100	40

2004-05	84	61	184	1406	23	8	7	863	5267	1875	123	40
2005-06	84	74	248	1146	15	11	11	795	5280	909	72	44
2006-07	119	40	227	1260	32	8	8	489	5856	0	61	44
2007-08	113	60	273	1464	24	19	18	2132	5189	0	118	44
2008-09	113	54	238	1454	27	38	26	4060	4112	1250	156	40
2009-10	113	50	184	894	18	16	17	1430	2977	0	84	50
2010-11	113	47	204	1028	22	19	19	1435	2880	0	76	46
2011-12	113	47	185	911	19	14	16	989	2687	0	62	45
2012-13	113	47	201	790	17	16	16	891	2147	0	56	55
2013-14	113	45	238	746	17	12	12	508	2484	417	42	60

In 2004, the Kibber insurance program was expanded to include livestock from three relatively smaller neighboring villages: Kee (35 families), Gete (6), and Tashigang (6). The number of participating families thus grew to 70, and the total insured livestock increased to 260 by 2006. Most members from Tashigang and Gete quit the program for a few years shortly thereafter, but rejoined later and are now participating in the program again.

Between 2002 and 2014, 182 large-bodied livestock insured under the Kibber program were lost to predation, while two died of disease. A total amount of c. US$ 15,370 was paid out from the insurance fund for these losses (see above, Table 11.2). The instances of disease-caused mortality of insured animals were treated as special cases, and half of the compensation amount was provided to the owners based on a collective agreement.

In 2005, two members of the insurance program made false claims regarding the loss of insured livestock to wild carnivores. Given the high community involvement in paying premium as well as managing the program, these false claims were easily detected; and the two were warned that such attempts would not be tolerated in the future.

By 2006, five years after its initiation, the Kibber insurance program became largely self-sustaining financially. Since then, further infusion of external funds has only taken place on two occasions, once when livestock losses in the previous year were exceptionally high (2008-09; 1250 USD), and another time in 2013-14 (420 USD), when the premium and compensation amounts were revised upwards for 2015 (Table 11.2).

A similar insurance program, adapted to local needs and conditions, was later started in the villages of Chichim and Sagnam in Spiti, increasing the total of participating villages in Spiti to six. The program was further adapted and expanded to Ladakh, another part of India's snow leopard habitat, to cover six villages (Rumtse Gya, Sangmath Gya, Rongthya Gya, Miru, Sasoma, and Kyungyam), largely focusing on horses.

The program has since been also expanded to the snow leopard habitat of South Gobi, Mongolia, and our colleagues and partners have recently initiated pilot livestock insurance programs in snow leopard habitats of China and Pakistan.

What it involves: steps in setting up a new insurance program

Setting up and running a livestock insurance program involves several steps, which are outlined below.

*Information is generated on extent and patterns of livestock depredation and socio-economy of affected peoples, livestock holdings, etc. (Appendix 11.1).
*Main communities in the landscape affected by livestock depredation are identified.
*More information is generated on circumstances of livestock predation, grazing and herd management practices, social structuring in the community, social capital, identification and contact with potential local champions and other knowledgeable people. All this information helps assess what kind of interventions might be useful, and whether an insurance program is potentially appropriate and feasible (see Chapter 3: APTNESS).
*Time and effort are invested in relationship building with the community (Chapter 2: PRESENCE).
*One on one discussions are conducted with a few knowledgeable people and local champions to share program details, gauge their interest, and assess local issues.
*Based on the above, at the appropriate time, community meetings are organized for information exchange and discussion on program details.
*After a series of such interactions, when the community appears ready, formal meetings are conducted to discuss program details and encourage the community to draft rules and regulations for the program. If a program is already

established elsewhere, it is useful to share its rules and regulations with the new community, while encouraging them to draft their own version.
*It is useful if an articulate insurance committee member or a local champion from another community can share their experiences about the program with the new community.
*Negotiations take place on premium and compensation amounts, and external funds contribution. Premium and compensation amounts are decided collectively for each age/sex class of livestock to be insured (Chapter 6: NEGOTIATIONS).
*Typically, the conservation agency provides 50-60% of the funds in the first 5 years, the remaining being contributed by participants through premiums. However, where communities are particularly poor, it is appropriate to increase external funds contribution above the typical 50-60 %, though one must keep in mind that this will increase the period and volume of external funds it takes for the program to become financially self-sustaining.
*It is important at this stage to help the community understand if some of their draft regulations are inappropriate, and to ensure that clauses that directly help wildlife and carnivore conservation are included in the rules and regulations (see below, Appendix 11.2).
*Through discussions, locally appropriate systems are devised for insurance committee functioning, discussing modality of premium collection, funds management, bank accounts, accounts books etc. A small amount is agreed upon to offset the costs (travel etc.) that the insurance committee incurs. That amount is usually paid annually from the insurance fund.
*A system for validation of claims by insurance committee members is established through community discussions, and timing of disbursement of claims (preferable once at the end of the year) is decided.
*Insurance committee is provided assistance and guidance to open bank accounts, maintain account books, etc.
*Insurance register includes the name of the owner, species, age, sex, and physical description of each animal insured. We have sometimes ear-tagged animals for identification, but this is not necessary if good descriptions on their appearance can be made in the register. The register also includes updated information on premium payments.
*When dealing with high livestock numbers, e.g. with goats and sheep, insurance can be done and premium payments be made at the herd level rather than for individual livestock (i.e. people insure all their goats/ sheep. In such a case, it is important that the entire herd is insured, and not just a part of it. Individual identification of livestock is not necessary in such cases.
*Premium collection is done either monthly or half yearly, or at any other suitable periodicity, though the premium payments must be made in advance rather than

retrospectively.

*Conservationists and insurance committee members maintain communication, and meet at least twice a year, to address any issues and to copy the data on animals insured, premiums paid, and any cases of livestock depredation.

*In the case of any livestock mortality for which compensation is to be claimed, the owner informs the insurance committee, whose members or their representatives conduct site visits and record all the details of livestock mortality, including, for each verified livestock kill, the details of the animal(s) killed (species, age-sex), carnivore signs recorded, location, date and circumstances of predator attack etc.

*Payments for lost livestock are made once a year at a community event after the insurance committee, conservationists, and village elders discuss and approve each claim.

*Each year, not more than 50% of the total insurance funds (accumulated premium and externally contributed funds) should be disbursed. Therefore, if in a given year, livestock mortality is high, every owner gets proportionately reduced compensation such that the total expenditure does not exceed 50% of the insurance fund. This system is important to ensure the sustainability of the program and must be included in the rules and regulations and explained clearly to the community.

*Once in a few years, through discussions with the participants, the premium and compensation rates are revised to keep up with changes in the market value of livestock. Additional infusion of funds is necessary for the program at this stage, and also serves as a negotiation tool to encourage the participants to increase premium amounts.

*Conservationists need to solicit and remain open to any new ideas that are suggested by the community. In Mongolia, one of the insurance groups started a microcredit program for its members to strengthen the insurance fund. In India, committees decided that in case of an insured livestock dying from disease (not old age), the owner would be eligible for half the compensation amount.

Specific good practices that help avoid problems:

*During meetings to discuss setting up of the program, if there is reluctance, it is important not to force a decision. Instead, it is useful to clarify doubts, and give people more time (till further meetings) to reflect.

*Premium amounts must be paid in advance, either monthly, twice or a few times a year, and not retrospectively.

*There are often potential participants who, despite having livestock, prefer to wait and watch, instead of joining the program initially. If they wish to join at a later date, a reasonable joining fee is levied, with the amount to be decided by the community. The community should be encouraged to make this fee neither too small (which would be unfair to the pioneers who pay premiums and help build the funds from the outset) nor too high (which would deter new members from joining). This should be recorded as a clause in the rules and regulations.

*As mentioned earlier, not more than 50% of insurance funds are ideally paid out each year.

*It is important to clarify and reiterate repeatedly that while all funds (premium and those provided by the conservation agency) will remain with the community, participants will not get refunds in case they don't lose livestock.

*Adding participants from a smaller community to the insurance program of a larger one should be avoided. To the extent possible, it is useful to start independent programs for each community.

*Before compensation is paid out at an annual public event, the insurance committee should present each claim for final approval to a group of village elders, ideally in the presence of a conservationist.

*Having a "good herder" prize (awarded for zero or minimal livestock losses to wild predators) encourages participants to better protect livestock. In case this is a monetary reward, the amount can be noted in the rules and regulations.

*It is important to remain flexible and jointly modify rules and regulations based on experience.

*It is helpful to raise awareness of the program amongst the larger community at least once a year in order to expand the program to cover more participants. It was quite helpful when we once offered to cover the joining fee of new participants during a membership drive.

* Organizing a fun event, e.g. a dance party, volleyball competition etc., after each insurance meeting can spread the message about the program to non-insurance groups, and helps team-building for insurance groups.

PARTNERS Principles and community-based livestock insurance

Community-based livestock insurance programs depend significantly on several of the PARTNERS Principles in order to be initiated and run successfully - more so than some of the other interventions in our conservation portfolio, such as SLE (Chapter 10) or corral improvement.

PRESENCE

An insurance program requires interested community members to make financial contributions in the form of premiums. Even though the money remains with the community, it moves from the individual to a common pool. Individuals in the community therefore must be able to trust the conservationists, and be convinced about their long-term engagement.

> Setting up an insurance program is unlikely to succeed if it is one of the first joint initiatives between the community and the conservationists and not based on a reasonable trust-building phase.

Keeping this in mind, one should hope to successfully set up a community-based insurance program only after a reasonably long-term presence and trust building with the community, and preferably after having run other conservation programs with them (Chapter 2: PRESENCE). Setting up an insurance program is unlikely to succeed if it is one of the first joint initiatives between the community and the conservationists and not based on a reasonable trust-building phase. There is an important role for local champions in helping set up insurance programs, as they can be a critical bridge between the community and the conservationist.

APTNESS

The threats and the science

Not all villages or communities living in a given snow leopard habitat are equally affected by livestock predation by large carnivores. Livestock predation varies over space (and time), and depends on several factors such as herding and guarding practices, the composition of livestock herds, the abundance of wild prey and carnivores, and where in the landscape livestock are grazed or penned (Suryawanshi et al. 2013, Johansson et al. 2015, Mishra et al. 2016b).

In some communities, the predominant problem involves predator attacks inside poorly constructed corrals, in which case corral improvement might be a more apt intervention, while in others, it could be livestock losses in the pastures, or both. Only for communities where livestock losses to predators are a serious and recurrent problem is a program like this going to be relevant.

> In some communities, the predominant problem involves predator attacks inside poorly constructed corrals, in which case corral improvement might be a more apt intervention, while in others, it could be livestock losses in the pastures, or both.

In order to assess the Aptness of insurance programs, it is essential to conduct extensive interview and discussion-based surveys of local people and key informants to understand the spatio-temporal extent of livestock predation (and other threats) before planning any intervention (Appendix 11.1). The same surveys can combine information gathering on the local socio-economy (that will determine the extent of people's ability to contribute premiums), skill sets, access to banks, grazing practices (that will enable an understanding of what could be done to reduce livestock mortality in the first place), and other causes of livestock mortality or livelihood challenges (that can provide insights into what other interventions could benefit people and conservation, e.g. a vaccination program; Chapter 12).

The scale

The number of participants in a community-based insurance program can significantly influence its outcome. A group that is too small is less likely to establish a sustainable program because of the difficulty in building a large-enough insurance fund, which is significantly dependent on premium amounts contributed by the participants. On the other hand, too large a group poses management challenges, increases the moral hazard and the tendency to make false depredation claims. Our partner communities that have successfully run insurance programs over the long term have ranged in size from about 15 to 80 families. If a community is 'too large', it might be best to divide it based on different pre-existing groups of families any such large community is likely to have (see Chapter 3: APTNESS).

Socio-economics and social capital

As mentioned earlier, the ability of participants to contribute premiums depends on their economic status; and in relatively poor communities, a greater proportion of the insurance fund may need to be subsidized by conservation funding.

The willingness of people to participate in a program such as this also depends on social capital. Much like the need for trust between the community and the conservationist, insurance programs rely on relatively high internal trust and social capital. They need systems and norms that can ensure, in this specific case, that the participants pay their premiums on time, and that the selected committee members manage the funds with integrity and transparency. They require social networks and trust to deal with the moral hazard, and to deter the

> Much like the need for trust between the community and the conservationist, insurance programs rely on relatively high internal trust and social capital.

temptation to file false claims. When such false claims are filed or when there are disagreements, they need to be able to resolve them amicably and fairly. They also depend on largely voluntary contributions of time and effort of insurance committee members to run the program. In cases where there are internal divisions in a community, or where necessary social networks and systems are weak, it may be best to invest first in other interventions that can help people and address threats to conservation while building social capital.

> Conservationists and the community must simultaneously explore ways to reduce livestock mortality in the first place through initiatives such as better herding or corral improvement. This is likely to help reduce the financial burden on the insurance program, reduce moral hazard for the insurance participants, and generally help in improving people's attitudes and behaviorst towards the predators.

Multi-faceted approach

As mentioned earlier, effective management of conflicts over livestock predation requires much more than the mechanism to share or offset economic losses that an insurance (or compensation) program is able to provide. Conservationists and the community must simultaneously explore ways to reduce livestock losses in the first place through initiatives such as better herding or corral improvement. This is likely to help reduce the financial burden on the insurance program, reduce moral hazard for the insurance participants, and generally help in improving peoples' attitudes and behaviors towards the predators. Conservation-linked livelihood programs (such as SLE, see Chapter 10) and sustained education and conservation awareness programs are also needed to help increase people's ability and tolerance for living with carnivores. It is also useful to keep in mind that an insurance program is unlikely to reach out to all families in any community. Some families may not have livestock, while others may opt to stay out for some reason, and therefore, multiple initiatives are desirable.

RESPECT AND EMPATHY

Respectful and empathetic engagement with the community is a necessary condition for any community-based conservation effort (see Chapters 4 and 7). Insurance programs need to be set up as equal partnerships. In the discussions, the autonomy of the community/participants must be respected, and their collective interests and concerns must be incorporated into the rules and regulations of the insurance program. A rather effective way we have found is to introduce the community to the concept and broad framework, and encourage them to come up with the draft rules and regulations.

TRANSPARENCY

Insurance programs are designed to assist the herders by sharing and offsetting wildlife-caused economic loss. It is important, however, that our main purpose of wildlife (and carnivore) conservation is clearly communicated to the community, and clauses that safeguard the well-being of wildlife are incorporated in the rules and regulations. Printed copies of the rules and regulations, once finalized, must be shared with all participants, while repeatedly reiterating the need for periodically reviewing and modifying them jointly.

Within the community, it is important to ensure transparency in the way premiums are collected, managed (ideally deposited in a local bank account opened specifically for the purpose), and account books maintained. Compensation amounts must be paid out in full view of the participants and the community, after reviewing each claim transparently.

We have found that people unfamiliar with the concept of insurance often tend to assume that should they not lose any livestock to wild predators, they can reclaim their premium contributions after a certain period of time, which is obviously not the case. Such doubts need to be clarified repeatedly, especially in the beginning.

NEGOTIATION

Given the strong need for trust and social capital, community-based livestock insurance programs like ours, that require the participants to contribute premiums, take sustained effort and engagement with the community before they can be implemented. As the community gets convinced about the value of such a program over time and develops a trustful relationship with the conservationist, the engagement slowly turns into a negotiation.

It is in the herder's interest, understandably, to be able to secure high compensation for livestock killed by carnivores while contributing relatively low premiums. This, however, would tend to make the program unsustainable over the long run, and a balance needs to be found between premium and compensation amounts. Making the amount of conservation funds to be provided contingent upon the amount of premium the community is willing to contribute is a useful way of negotiating this balance. Typically, we offer to match (or even marginally exceed) the amount of premium the participants are willing to contribute with conservation funds for at least the first 5 years. Market value of livestock provides a fair standard upon which to base these negotiations (see Chapter 6: NEGOTIATION).

As mentioned earlier, the situation in every community is unique. The approach must therefore be flexible, as communities live in different conditions, and have different concerns and interests that need to be incorporated into the insurance program. For instance, we have a few communities that insure only horses, others that insure all large-bodied livestock, and still others that insure all large and small livestock. Their premium and compensation amounts also vary. This means that the rules and regulations often need to be tweaked at the level of each community or group, so every insurance program could vary from the other in terms of the livestock insured, premium contributions made, or the compensation paid. As we will see shortly, flexibility in approach is not just needed across communities, but also over time.

The negotiations must also lead to incorporating the needs of wildlife conservation into the insurance rules and regulations, including clauses for no persecution of wildlife, leaving behind and not retrieving carcasses when livestock is killed in the pastures away from settlements, etc.

It is important to record the agreements, rules and regulations formally, and for copies to be shared with all participants. This reduces the chances of confusion, and allows for more amicable resolution of disagreements.

Finally, it is important in the negotiations (and subsequently) that the program is projected as the community's rather than that of the conservation agency. Indeed, in the case of community-based livestock insurance, participants bear most of the cost, typically contributing 40-50% of the total insurance fund through premiums; and they largely run the program themselves. Negotiations must reiterate and ensure a high level of ownership and pride of the participants in the program.

RESPONSIVENESS

It can sometimes take months or even years of relationship building and information exchange to garner the level of trust and social capital needed before a community shows the interest and willingness to pilot a self-managed insurance program. This is necessarily a slow process, but one must be responsive and move quickly when a community begins to show serious interest. Quick progress is possible. Yet, sometimes, despite showing interest, community-members are reluctant to actually start the program, in which case, time, patience and continued engagement are required.

While developing and finalizing the rules and regulations of any insurance program, and indeed at any subsequent opportunity, it is important to keep reiterating to the participants that the program will evolve as we jointly learn from experience. The rules and regulations will need to be modified, including premium rates, compensation rates, etc., though any change would be based on full participant engagement, discussion, and mutual consent of participants and conservationists.

Conservationists must maintain continued and strong communication with the committee members and participants. We have found it useful to conduct at least two formal meetings with the insurance committee each year. This helps ensure that things are proceeding smoothly, helps detect any mistakes or challenges before it is too late, and provides an opportunity to copy the insurance data for long term monitoring.

Conservationists must maintain continued and strong communication with the committee members and participants. We have found it useful to conduct at least two formal meetings with the insurance committee each year.

Even after an insurance program has become self-sustaining and runs entirely on premium contributions of participants, these meetings of engagement are important. The conservationists also need to be responsive and bring infusion of funds into the program during the odd years when there is exceptionally high livestock damage. External funds infusion is also appropriate when, once in a few years, the compensation and premium amounts are being renegotiated to respond to changing market value of livestock.

STRATEGIC SUPPORT

As mentioned earlier, many governments have tried to and others continue to run compensation programs with various levels of success. There is the opportunity here to influence government policy and encourage support for community-based insurance programs rather than the continuation of state-run compensation programs, which even governments have found difficult to manage over the long-term because of a variety of reasons discussed earlier.

Appendix 11.1

General framework for survey and situation analysis to assess the need and potential of community-based livestock insurance programs

This is not comprehensive, but indicative of what kind of information would be useful. Survey sheets should be designed for individual interviews and group discussions.

1. Names of village
2. GPS Location
3. Detailed survey of livestock abundance in the village
4. List of primary sources of income with a rough division of earning (Livestock, agriculture, tourism etc.)
5. Pair-wise ranking of causes of livestock mortality
6. If predation by carnivores is listed:
    a. Species of carnivores involved
    b. Number of incidences of predation in the past one year
    c. List the following for each instance of predation
        i. Time of year
        ii. Location (corral or pasture: where in pastures)
        iii. Species and age of livestock
        iv. Time of day
        v. Economic value of the livestock species
7. What are the primary crops grown in the village?
8. Primary problems for agriculture in the village?
9. Primary challenges for livestock rearing?
10. Resource map of the village (to be mapped):
    a. Livestock movement across seasons (which pastures are utilized)
    b. Description of other problems of livestock mortality
    c. Carnivore occurrences (which are the areas of usual carnivore sightings)
    d. Wildlife patterns (areas and seasons of wildlife sightings around the village)
    e. Crop damage (if it comes up in the discussion then mapping of areas)
    f. Corrals (mapping of night time corrals if livestock is penned away from the village)
    g. Others things of significance (e.g. description of other sources of income)

Appendix 11.2

General framework for agreements between the community and conservationists while setting up community-based livestock insurance programs

Insurance committee:

- The participants of the insurance program designate a committee responsible for collecting premiums, managing all monetary transactions and maintaining insurance registers. They may choose to rotate this responsibility after a suitable amount of time.
- The committee validates insurance claims.
- Committee members receive a nominal honorarium from the insurance fund to take care of transport and other program-related costs.
- At least two meetings are conducted every year between the insurance committee and the conservationists.

Insurance fund:

- During the initial years (usually the first 5), approximately 60% of insurance fund are to be provided through conservation funding, while the remaining is to be collected by participants in the form of premiums. Annual external contribution for the insurance fund will stop after this initial period, but the partner organization is still available to provide assistance in future years if and when needed.
- Premium is preferably collected for the whole year in one or two installments, but participants may also choose to make monthly payments. All premium payments must be made in advance (at the beginning of the month or year as appropriate).
- If a participant pays premium regularly for a certain number of years (3-5 years) and does not claim payment for any livestock mortality, the following year's participation in the insurance program can be made free (no premium to be paid for that one year).
- Premium is not returned if the owner sells his/her livestock after having paid for a whole year.

Compensation distribution:

- Only 50-60% of the money available in the insurance fund is distributed as compensation in any year. If there are too many claims, the compensation amount per livestock is to be proportionately and equitably reduced.

- Compensation is to be paid once every year, usually before collecting premium for the next year.
- Compensation is provided, in good faith, in instances where determining the exact cause of death or disappearance of livestock is difficult.
- Sale of insured livestock within the same village can lead to transfer of compensation to the new owner if the premium for the year has been paid.
- While giving out compensation, a large proportion of the members should be present along with one representative from the partner organization.
- No compensation is given and the premium paid is not returned for livestock that are killed for slaughter or for any other reason by the owner.

Joining Fee:

- Any community member who owns livestock when the program is initiated but chooses to join later pays a joining fee, the amount for which is collectively decided.
- Any community member who starts rearing livestock after the program has been initiated and wishes to join it does not have to pay a fee, or a nominal fee which is smaller than if he had owned livestock before the insurance program's start date.

Herding:

- A monetary reward can be instituted for herders under whose watch the least number of depredation cases are reported.
- Funds from the insurance corpus can be used, with consent from all participants, to hire extra herders during the busy agricultural season.

Wildlife:

- Any animal killed in the pasture is not brought back to the village for meat or any other reason. The carcass is left in the pasture.
- No persecution of snow leopards, their prey, or other wildlife is allowed.

Roles and responsibilities of key stakeholders

Participant community takes the responsibility for the running of the program:

- They elect a committee and see to its functioning.
- They pay premium at the agreed time.
- No false claims are to be made.
- There must not be any retaliatory killing or poaching of wildlife. Any such incidents are to be reported immediately.

- They must improve their herding measures to protect loss of livestock.

The committee is responsible for the smooth functioning of the livestock insurance program:

- They collect premium at the agreed time.
- They open a joint bank account and deposit all collected money in it.
- They must make sure that compensation is paid to only valid cases.
- They maintain a register with all relevant details such as total number of livestock, amount of premium paid, and funds available in the bank, identification marks on livestock.
- They gather data on depredation.
- They make sure all transactions carried out are fair and transparent.

Conservation organization:

- The partner organization pays up to 60% of the total insurance fund during the first five years.
- Representatives attend meetings and are available to help and support the running of the insurance program.
- They collect data, analyze it and provide information for program improvement.

Chapter 12:
SETTING UP COMMUNITY-BASED LIVESTOCK VACCINATION INITIATIVES FOR WILDLIFE CONSERVATION: A CASE STUDY OF ECOSYSTEM HEALTH PROGRAM

Contributions: Muhammad Ali Nawaz (text and information)[3]

Predation on livestock by large carnivores is only one of the several challenges that farmers face while trying to make a living from livestock rearing in wildlife habitats. Disease is another such challenge. In many snow leopard landscapes, for example, adequate veterinary support is unavailable to farmers. Livestock losses to disease can be substantial, sometimes even higher than the losses to wild carnivores. In northern Pakistan, for example, our surveys estimated that annually, 3 to 14 % of the livestock holding was lost to diseases in different valleys (Snow Leopard Foundation, unpublished data).

Disease in livestock also poses a high risk of disease outbreaks in wild ungulate prey of snow leopards, given the considerable overlap in habitats, plants eaten (Bagchi et al. 2004; Mishra et al. 2004), and water sources used by livestock and wild ungulates. An outbreak of *Mycoplasma capricolum* pneumonia in 2010 that is thought to have killed 20% of the population of the endangered markhor

[3]*Nawaz, M.A. and Mishra C. (2016). Setting up community-based livestock vaccination initiatives for wildlife conservation: a case study of Ecosystem Health Program. Pp. 149 to 162 in Mishra, C. (Ed.) The Partners Principles for community-based conservation. Snow Leopard Trust, Seattle, USA.*

> Disease among livestock also poses a high risk of disease outbreaks in wild ungulate prey of snow leopards.

Capra falconeri in Tajikistan is believed to have originated from domestic goats (Ostrowski et al. 2011). A fatal outbreak of scabies in blue sheep *Pseudois nayaur* in the Pakistan Pamirs is suspected to have been transmitted from livestock (Dagleish et al. 2007). Another outbreak of PPRV in the wild goat *Capra aegagrus* in Pakistan was fatal, but appeared to be controlled in those areas where livestock in the surrounding villages had been vaccinated (Abubakar et al. 2011).

Evolutionarily, wild ungulate prey of snow leopards, inhabiting the cold South and Central Asian mountains, have perhaps been less exposed to pathogens and parasites compared to livestock, and may therefore be less immune and more susceptible to disease transmission from livestock (Dagleish et al. 2007).

Our Ecosystem Health Program (EHP) is a response to the economic losses farmers in snow leopard habitats face due to livestock diseases and to the potential transmission of diseases to wild ungulates. We assist farmer communities in establishing a sustainable livestock vaccination program that enables them to reduce livestock mortality and strengthen their livestock-based income.

The participating farmers commit to a ceiling on their herd size as well as a ban on hunting snow leopards and their prey. Like SLE (Chapter 10), the EHP is a conservation-linked incentive program that aims to indirectly increase the ability of farmers to coexist with snow leopards. It encourages, trains and assists farmers in better veterinary care and vaccination of livestock, thereby also reducing the chances of disease transmissions from livestock to wild ungulates.

EHP is especially relevant in those snow leopard areas where farmers do not have access to appropriate veterinary care for their livestock. The program involves increasing the awareness of livestock health issues among local people, training community representatives in livestock healthcare and vaccination, and helping create a community fund for the purchase of vaccines.

Program in operation

Our surveys estimated average annual livestock losses to disease in some of the snow leopard habitats of Pakistan to be 7.8% of livestock holdings (Snow Leopard Foundation, unpublished data). This amounted to an estimated 60% reduction in disposable income for the affected families. It also presumably affected the conservation status of snow leopards directly, as some fatal disease outbreaks in wild ungulates in Pakistan are thought to have been linked to livestock (see above).

To assist local communities and improve conditions for snow leopard conservation, EHP was initiated in 2003 in the village Kuju of Chitral District in Khyber Pakhtunkhwa, and expanded to Parsan in 2005. This was initially done in partnership with WWF Pakistan, where SLT's program was based until the Snow Leopard Foundation Pakistan was established.

Four villages (Mori Payeen, Koghozi, Barkhozi, Bakhtoli) were included in 2009, and another five villages (Balim, Drungagh, Rech, Sor Laspur, Ujnu) of Chitral District joined in 2010. The program was subsequently expanded to six valleys in Gilgit Baltistan (Qurumbar, Phandar, Rakaposhi, Hispar, Shimshal, and Chuparson).

The program in Kuju started in 2003 in one (Kuju Peyen) of the two hamlets comprising the village. The livestock holding of Kuju Peyen had been around 1500 animals (800 goats, 100 sheep, and 600 cattle) in 2003, but it declined to 1050 by 2008, and further down to 739 (545 goats, 13 sheep, and 181 cattle) by 2014. This decline is consistent with a general pattern in northern Pakistan (Livestock Census, 2006), and is probably linked to young people increasingly preferring other jobs to livestock rearing. Many young people are getting opportunities in the government and private sector within the provinces or in big cities further afield. The costs of food and other commodities have also increased in recent years, compelling some people to sell livestock to meet their financial needs.

An estimated 100 livestock had died of disease in Kuju Peyen in 2002, the year before EHP was initiated. In 2007, in the fourth year of the program, c. 10 goats and 5 cattle were reported to have died of disease.

The average cost of the vaccination program in Kuju Peyen was PKR 60,625 (US$ 596) per year. An estimated 100 livestock had died of disease in Kuju Peyen in 2002, the year before EHP was initiated. In 2007, in the fourth year of the program, c. 10 goats and 5 cattle were reported to have died of disease. During the same year, c. 10-15 goats and 10 sheep were reported to have been killed by snow leopards. Six people from Kuju Payen have been trained over three training programs as Ecosystem Health Workers (EHW), and four out of them are actively working in the program at present.

The program in Kuju Peyen continues to the present day, and our staff continues to be responsible for providing vaccines, monitoring and oversight. No instances of hunting of snow leopards or any illegal hunting of wild ungulates have been recorded since its inception. While there is little doubt about the positive impact of the program for conservation, we do believe that the program currently has an inadequate ability to detect potential violations of the conservation contract. We are trying to improve the monitoring system of the program.

In the first year, we provide the entire cost of vaccines, while participants contribute into a central fund for future vaccines. In the second year, we cover 75% of the cost of vaccines and contribute the remaining 25% to strengthen the vaccination fund. The participants contribute 25% of the vaccine cost. In the third year, we provide 50% of the cost of vaccines and contribute the same amount into the vaccination fund, while the participants contribute 50% of the vaccination cost. In the fourth year, our entire contribution goes into the vaccination fund, while the entire cost of vaccines is borne by the participants. After the fourth year, our contribution for the purchase of vaccines and for the vaccination fund is stopped, and the full cost of vaccines and vaccination is borne by the participants.

In 2014, an evaluation survey across a range of communities including Kuju Peyen showed a reduction in disease-caused mortality of livestock of 50% (2.2 to 1.06 livestock heads per family annually) in EHP areas. On average, 83% of the total livestock were vaccinated under the program. A majority of the respondents (68 %) perceived a positive change in livestock health, with increased weight gain, draught power, and milk yield, while the remaining 32% did not perceive any change.

The surveys estimated that 1.84 livestock heads per family are currently sold each year. Additionally, 1.45 animals per family are used annually for personal consumption.

What it involves: steps in setting up EHP

Our program is currently running in 18 communities in Pakistan. The general steps employed are as follows:

*Community-level information is generated on the key challenges to livestock, relative importance of disease as a cause of livestock mortality, importance of the area for wildlife, the extent of overlap between pastures and snow leopard habitat, local socio-economy, herd size etc.

*Regional information is collected from literature and the veterinary departments on the main livestock diseases prevalent.

*Interactions with key informants and local champions are held to explain the program, and based on the community's need and willingness; the potential for program implementation is assessed.

*If the conditions appear suitable based on the information generated and feedback received from community representatives, formal meetings with the larger community take place where the program details are explained. If the community is keen to participate, they are encouraged to designate a representative Snow Leopard Conservation Organization (SLCO); a community-based organization to oversee the program. SLCOs are formed with participation of the community, and led by elected community members.

*Based on the main livestock diseases, a vaccination schedule is prepared.

*A conservation agreement is developed and signed by the community (represented by SLCO) and the conservation agency (Appendix 12.1).

*The community designates the members to be trained as Ecosystem Health Workers (EHW). We try to train at least two members from each community, and more if the budget allows. The number of participant families per EHW currently ranges from 2 to 8, and the number of livestock vaccinated by an EHW ranges from 2000 to 3000. From our experience so far, we believe that each community should have at least two Ecosystem Health Workers and an ideal ratio of one EHW for every 2000 to 3000 livestock for efficient functioning of the program.

*The trainings for EHWs are arranged at a central location, such as a veterinary institute or university, and relevant experts are involved or hired. The training focuses on enhancing the understanding of livestock healthcare and diseases (Appendix 12.2), and enables trainees to conduct vaccination according to the vaccination schedule. The training sessions for new EHWs typically last 7-10 days if organized in the program sites, or four weeks if organized at a professional institute.

*The quantity and type of vaccines needed is assessed based on the vaccination schedule and in consultation with the local officials of the Government Livestock Departments. A locally adapted and efficient vaccine procurement and delivery system is created. Vaccination campaigns take place twice a year, in spring and fall.

*The vaccines are purchased from dependable sources such as Veterinary Institutes or other reliable agencies to ensure quality and validity, and are usually available at a cost that is subsidized for the farmers by the Government.

*The vaccines, along with forms to record vaccinated animals (see below, Appendix 12.3), are given to the EHWs. The EHWs return the completed forms to our staff after each vaccination campaign. Once a year (during the fall vaccination campaign), EHWs collect information on livestock holdings and fate of livestock during the previous year, including mortality and its causes, sale and sale price etc.

*The EHWs are considered self-employed, and receive a small amount of money (currently PKR 5 or US$ 0.04 per livestock per session, or US$ 0.09 per livestock per year) for their service from each participating household. The amount is decided collectively by the participants and the conservation agency and periodically revised to keep up with prevailing costs.

*During the vaccination session, each household currently also contributes an additional PKR 10 (US$ 0.09) per livestock to the community-level vaccination fund for procurement of the next round of vaccines. This is the amount at which a round of vaccines is currently available at subsidized rates.

*A fund is created in each community and serves two purposes. It acts as a catalyst for encouraging the farmers to pay for the cost of the vaccines and formalizes the system of payment, vaccine procurement, and vaccination. It can also provide a small amount of buffer, so that the program can survive minor emergencies, such as an unexpected disease outbreak or a rise in vaccine costs.

* As mentioned earlier, In the first year, 100% of the cost of vaccines is borne by the conservation agency, while participants start making their contributions into the fund for future vaccines. In the second year, the conservation agency covers 75% of the cost of vaccines and provides the remaining 25% to strengthen the vaccination fund, while 25% of the vaccine cost is borne by the participants. In the third year, 50% of the cost of vaccines is provided by the conservation agency and 50% goes into the vaccination fund, with the participants paying for half the cost of vaccines that year. In the fourth year, the entire amount goes into the vaccination fund, while the entire cost of vaccines is borne by the participants. After the fourth year, external contribution for the purchase of vaccines and for the vaccination fund is stopped, and the full cost of vaccines and vaccination is borne by the participants.

*The program is monitored twice a year by our staff together with the local Wildlife Department and SLCOs. Monitoring entails reviewing vaccine administration, its impacts on livestock health and well-being of the community, and assessing conservation compliance.

*Conservation compliance is monitored through maintaining annual records of livestock holding per family, extent of their grazing areas, and incidence of any wildlife persecution in the community.

Specific good practices that help avoid problems:

*The rights and responsibilities of SLCOs must be agreed upon and periodically reiterated.

*Continuous informal interaction (or at least once a month on average) with local champions and representatives of the SLCOs is essential to ensure resilience and efficiency of the program.

*A strong education component that reiterates the conservation connection of the program is important to ensure that the program does not come to be seen purely as an animal husbandry initiative.

*EHWs play a critical role in the program. They must be selected by the community (see Chapter 5: TRANSPARENCY). We have found that candidates who are not educated at least till high school often get inadequately trained. On the other hand, those who are well educated (graduation and higher) are often in pursuit of external jobs and tend not to stay in the community very long. It is therefore advisable that the qualification criteria for EHWs shared with the community or SLCO specify a minimum level of education (around high school), experience with and knowledge of livestock, and high likelihood of staying on in the community.

*With proper training, EHWs become highly committed workers with a high degree of cost-effectiveness, and are motivated to convince the majority of livestock owners to vaccinate their herds. Over the course of the program, the role of the EHWs can be expanded to provide first aid services, manage reproduction disorders, and treat injuries as well as common diseases. With better professional training, EHWs can become adequately self-employed, and remote communities are able to get quality veterinary services at their doorstep.

*While the program is designed such that no external funds are provided after 4 years, the conservation agency must assure continued support and come in with additional contributions in cases of emergencies such as large disease outbreaks.

*The conservation agency must continue its role as a conduit for supplying vaccines, program monitoring, conservation education, and interacting with participants and local champions even after 4 years, when the external funding for the program is phased out. Without continued involvement, the Ecosystem Health Program can come to seen solely as an animal husbandry project and lose its conservation connection.

*We have found it useful to select at least 10 households randomly after each vaccination campaign to cross check the number of livestock vaccinated by the EHWs and quality of service delivered. This can be combined with general interaction with participants and non-participants to obtain their feedback on the program.

PARTNERS Principles and the Ecosystem Health Program

EHP can bring direct, quantifiable benefits to the participants over a reasonably short duration of time. Like SLE (Chapter 10), it can serve as a long-term conservation engagement initiative as well as a useful community entry tool. Similarly, like with SLE, the conditions for initiating EHP are less restrictive from the perspective of the PARTNERS Principles, though principles such as Respect, Empathy and Responsiveness have universal importance.

EHP can help make the partnership between the community and the conservation agency stronger, and help strengthen several of the PARTNERS Principles on which any such relationship must be founded. However, unlike SLE, which has general applicability in a wide range of conditions, EHP is appropriate specifically for situations where livestock disease is an important issue and farmers do not have access to quality veterinary healthcare.

PRESENCE

Communities that face serious economic challenges from livestock diseases and do not have access to adequate veterinary services can be expected to show ready willingness to participate in EHP. To that extent, the initial need for immersion in the community is largely related to understanding the Aptness of the initiative and building relationships with the community, especially the community champions. However, presence and immersion are necessary for sustaining the program, and for ensuring that it continues to have a strong conservation dimension. In the absence of continued interaction with the participants, the SLCO, and the community at large, EHP can easily come to be viewed as an animal husbandry initiative, and the conservation connection can get lost.

APTNESS

The threats and the science

EHP, as mentioned earlier, is appropriate for communities who face problems with livestock diseases and don't have access to quality veterinary healthcare. It is unlikely to be embraced by communities who already have access to veterinary services, often provided by Government departments in snow leopard habitats. In some areas, Government departments already conduct vaccination drives. It is therefore important that extensive community-level information is collected on the challenges faced in livestock rearing, prevalence of livestock diseases, and the levels of current access to vaccines and veterinary healthcare.

The scale

From a conservation perspective, it is important that the majority of livestock owning families in any community, and a majority of communities in any snow leopard landscape, participate in EHP or any other livestock vaccination program. Between EHP and other conservation initiatives, conservationists must try for almost complete participation of families. Further, specifically in the case of EHP, unless the vaccination coverage of livestock in the program is near complete, it would be difficult to prevent disease outbreaks, especially transmission to wild ungulates. While EHP is implemented at the level of the community, it is important to have a landscape-level approach. Targeting a majority of livestock, which might belong to different communities in any snow leopard habitat, is important for efficient control of diseases in livestock and transmission to wild ungulates.

Values

EHP can really benefit families and communities by creating disposable income, reducing the uncertainty in and improving livestock production. To increase the direct involvement of women, we have recently started to encourage the nomination of women to be trained as EHWs. So far, eleven women from four communities have been trained. Seven of them continue as active EHWs.

Socio-economics and social capital

We have been fortunate to have access to vaccines that are subsidized by the Government in Pakistan. Had that not been the case, one would have needed to work out a different financial model for EHP, especially for less affluent communities. We would have needed to subsidize the cost, and to provide funds for a much longer period or even into perpetuity, while having the participants share some of the cost.

As mentioned earlier, wherever livestock is an important source of livelihood and disease is an important cause of livestock mortality with little access to veterinary services, EHP would be effective. Like SLE, initiating EHP is less dependent on initial social capital. It helps create such social capital by bringing people together through the community fund, helps develop a trained cadre of community representatives in livestock healthcare, helps combine traditional knowledge with scientific techniques for livestock disease management, and creates linkages of the community with external institutions.

Multifaceted approach

EHP is a useful program for communities suffering from livestock diseases, and it helps strengthen livelihoods. It also reduces the risk of disease transmission to wild ungulates. However, from the perspective of snow leopard conservation, by itself, EHP is insufficient. EHP is an indirect way of improving conditions for conservation within communities. Even for addressing the threat of retaliatory killing of snow leopards, contextually appropriate initiatives such as livestock insurance and corral improvement will enhance the effectiveness of EHP when supported by good conservation awareness initiatives.

TRANSPARENCY

For the EHP to serve as an effective conservation tool, it's essential that we clarify our conservation goals at the outset and reiterate them periodically. Otherwise, the program faces the risk of coming to be viewed as a livestock husbandry project with no conservation connection.

While the desired qualifications for EHWs (high school education, knowledge of livestock, likelihood of continuing to live in the community) should be clearly communicated to the community representatives by the conservation agency, it is important that the selection of community members to be trained as EHW is made by the community in a transparent manner.

NEGOTIATION

That a community would try to negotiate the terms of any collaborative initiative is understandable, and is actually useful as it usually helps improve the terms of agreement and increase the community's ownership over the program (see Chapter 6: NEGOTIATION). Where Government subsidized vaccines are available, such as in Pakistan, the arrangement for cost sharing is more straightforward. However, there would be greater need for negotiation, in the interest of program sustainability and practicality, if an alternate cost sharing model needed to be worked out in the absence of Government subsidies. Conservationists must remain open to such negotiations,

> Conservationists must remain open to negotiations, and use objective standards such as the cost of vaccines and the economic status of a majority of the participants to negotiate the terms of cost sharing.

and use objective standards such as the cost of vaccines and the economic status of a majority of the participants to negotiate the terms of cost sharing.

Similarly, the participants must be encouraged to decide on the payment to EHWs based on objective standards such as the cost of skilled labor, the total number of livestock likely to be vaccinated by an EHW, etc.

RESPONSIVENESS

Currently, one of the potentially weak areas in EHP is the lack of a built-in system of rewards for encouraging conservation compliance, comparable to the system SLE employs (see Chapter 10). One of our initial thoughts was to stop working with a community if any incident of wildlife persecution did take place. However, we realized that doing so would neither be fair nor serve the purpose. Walking away is not an option in community-based conservation (see Chapter 5: NEGOTIATION). It is therefore important that a mechanism that rewards compliance and discourages non-compliance is built into the program. To this end, we are considering a system where a bonus is paid into the vaccination fund annually for conservation compliance, even after the fourth year. In the case of minor violations of the conservation contract, a proportion of the bonus would be withheld, while a community would lose the entire bonus for the year if there were a major violation. Additionally, we are exploring ways in which we can improve the systems for assessing conservation compliance.

In our model of EHP in Pakistan, the need to provide funds externally for purchasing vaccines or for strengthening the vaccination fund goes away after four years (this would change if we were to institute a bonus system). However, the conservation agency must remain ready to support the cost of vaccines even after four years, in cases of emergencies such as a major disease outbreak.

STRATEGIC SUPPORT

Involvement of Government departments in EHP is useful. In Pakistan, we work with Government Veterinary Institutes, who deliver training to EHWs and provide vaccines at subsidized rates. Indeed, Governments tend to have vaccination and veterinary care programs for local communities in most areas. However, these are often not put to efficient use. Partnerships like EHP can assist the Government departments in delivering quality service, can help communities, and can help conservation.

Appendix 12.1

Key points covered in conservation agreements between local communities and the conservation agency in the Ecosystem Health Program

Responsibilities of the local community, represented by the Snow Leopard Conservation Organization:

- Selecting community representatives (potential Ecosystem Health Workers) to get trained in livestock healthcare and to administer vaccines.
- Paying the community share of the vaccine cost.
- Keeping the herd size constant by encouraging the disposal of excess animals.
- Documenting snow leopard evidences and wild ungulate sightings, and predation reports.
- Protecting snow leopards and wild ungulates.

Responsibilities of the Conservation Agency:

- Organizing training workshops for the selected Ecosystem Health Workers to administer vaccines and assist in general livestock healthcare.
- Developing a locally appropriate vaccination schedule.
- Facilitating the access to reliable vaccines at subsidized rates.
- Monitoring the program and its socio-economic and ecological impacts.

Appendix 12.2

Content of the training courses for Ecosystem Health Workers.

The training courses contain a combination of classroom lectures and practical sessions.

The lectures provide a basic introduction to:

- Body organs and function
- Signs of good health and illness
- Diseases in livestock, causes and transmission
- Use and handling of drugs
- Use of traditional medicines
- Farming principles and importance of balanced nutrition

The practical sessions cover:

- Restraining animals

- Clinical examination and diagnosis
- Use of thermometer and treatments
- Treatment/control of ecto- and endo-parasites
- Use and handling of drugs
- Vaccination administration

Appendix 12.3 Key points covered in the vaccination forms

- Village Name
- Valley Name
- Name of the head of household
- Number of livestock of each type vaccinated
- Number of livestock of each type lost to disease during the previous time period
- Number of livestock of each type lost to different predators during the previous time period
- Number of livestock of each type sold and amounts
- Number of livestock of each type slaughtered for consumption

PART III.
CONCLUDING THOUGHTS

Chapter 13:
CONCLUDING THOUGHTS ON PARTNERS PRINCIPLES

In this book, I have outlined the main practical and ethical issues I believe are important to consider when engaging in community-based conservation. Chapters 2 to 9 each were devoted to one principle. In Chapters 10-12, my colleagues and I described, from a practitioners' perspective, three of our several community-based initiatives. Future versions of this book will hopefully carry more such descriptions of our other initiatives.

It is important to keep in mind that there is no one way to do community-based conservation, and the PARTNERS Principles are not a step-by-step guide for how to do it. They constitute a resource, a collection of ideas that must be given consideration while engaging with community-based conservation. The PARTNERS Principles could be used by a practitioner to chalk out steps to be taken in community engagement, or by a granting agency for informing its evaluation system.

I have argued that effective and sustainable conservation requires a fundamental shift in the way local communities have generally been viewed: from an undesirable presence or even a perpetrator to a real partner in conservation–despite their conservation-unfriendly activities, and despite the degradation their unsustainable natural resource use may cause. As I have mentioned, the degradation of natural resources due to unsustainable harvest by local communities is often an easier and more manageable problem to address compared to the drastic impact of and habitat loss due to large-scale commercial and infrastructural projects today.

Conservation needs friends. I have discussed in detail the role of conservationists, local people, and governments in community-based conservation. One group I have left out deliberately is the industry, the main agent of economic growth and the main vehicle, if not the cause, of biodiversity loss.

This omission should be viewed as a lacuna in this book, and not taken to imply that it is unimportant. I have desisted from invoking its role because we have little experience so far in engaging with the industry for conservation. This is poised to change. One of our new pilot programs aims at promoting snow leopard conservation by bringing together the industry and the Cashmere goat farmers in snow leopard habitats. We hope to catalyze a system and market for 'snow leopard friendly cashmere' that could assist in conservation as well as help increase revenues for the farmer and the industry.

The strong need to engage with the industry notwithstanding, local communities will remain biodiversity conservation's most important potential ally. It is essential that we start generating more support for conservation among them than we have so far managed or attempted to. Indeed, we need to start viewing the willingness of local communities to participate in conservation with appreciation and gratefulness, especially considering the tradition of top-down conservation that has historically tended to marginalize them.

The challenge, of course, is to enable such conservation partnerships and positive conservation action. As we have seen, building mechanisms for sustainable conservation is a constant and complex effort. That is what community-based conservation is about.

Who can do community-based conservation?

Biodiversity conservation is a societal obligation, as is working with under-served people. Community-based conservation offers an opportunity to do both and can be a fulfilling pursuit.

Imagining community-based conservation, however, is easier than achieving it. As we have seen, effective and respectful engagement, which is fundamental to the success of community-based efforts, is not just reflected in our conduct and civility of interactions with local communities. There is a potentially important role of our internal psychological orientation in the way we engage with local people.

It would therefore be unreasonable to expect every wildlife manager, trained in a hierarchical top-down approach to conservation management, to be able to effectively engage with communities. It is similarly unreasonable to expect many conservationists to be able to or to want to do so, considering that conservation has so far largely remained a pursuit of the privileged.

We are still a long way from the dominant conservation paradigm to create a central role for local communities. But there are many exceptions, and that's a start. Over time, the value of community engagement will hopefully get engrained into conservation systems, training, and thought. But until then, well-oriented conservationists must assist wildlife managers in community engagement.

Exit strategy

When would we have succeeded? I have been asked this question occasionally, have thought about it a lot, but have so far failed to find the answer. Maybe there isn't one, because it really means achieving the seemingly impossible situation where biodiversity conservation becomes so successful it no longer requires attention.

We would definitely have come a long way if and when biodiversity conservation no longer remains predominantly a pursuit of the privileged. When pro-conservation behavior becomes a part of life of the individual, and of decision-making systems of governments and local communities alike. When conservation becomes a standard part of the welfare state. If it is no longer considered a burden, or a hindrance to economic growth that must always be compromised, or somehow accommodated.

But as I have written earlier, there are no exit strategies in community-based conservation. In my experiences as both a grantee and an advisor to funding agencies, I have been struck by this apparent contradiction between the spirit of the PARTNERS Principles and the real world of conservation funding, where exit strategies tend to be emphasized considerably. It is useful to briefly discuss this seeming disconnect.

> The potential of an intervention to acquire financial self-sustainability is indeed a useful metric to evaluate its potential, but it should not be the only one, or even the most important one.

Funding agencies, like any other institution, have the right to decide on the nature and scope of their funding. Many, understandably, view themselves as catalysts and enablers, rather than sources of long-term funding for specific species, areas, or problems. Similarly, the desire of funding agencies for resources to enable biodiversity conservation to be used efficiently and for an adequate social return on their financial investment is not just understandable, but essential.

This premise often implies the legitimate expectation that, over time, the need for financially supporting an intervention should be reduced, and finally eliminated, except for supporting occasional and needed improvements. The potential of an intervention to acquire financial self-sustainability is indeed a useful metric to evaluate its potential, but it should not be the only metric, or even the most important one.

While some interventions such as SLE (Chapter 10) and livestock insurance programs (Chapter 11) are designed to achieve financial sustainability, others such as village reserves or community-led protection or monitoring, will require sustained financial inputs - either from funding agencies and donors, or from the government. As we have discussed before, each intervention tends to address a different issue, a different threat; and a multi-pronged portfolio is essential for effective community-based conservation.

On the one hand, therefore, the practitioners must try hard to build levels of sustainability into their interventions. The funding agency, on the other hand, can help ensure that the balance in the intervention portfolio does not tilt disproportionately toward certain interventions largely because they promise greater financial sustainability.

Too much focus on financial exit strategies can be counter-productive for community-based efforts because biodiversity conservation is not about building wealth, or even social capital. Community-based conservation requires long-term presence and management into perpetuity, just like investing into a Protected Area. An effective conservation portfolio will usually include relatively self-sustaining interventions as well as those that will need to be supported into perpetuity. As I have discussed earlier, the conservationist needs to be prepared for the long haul. So do funding agencies and governments who support these efforts.

More on success and failure

Conservation, as I have discussed, can be considered a wicked problem. There is no correct solution. Sometimes, we get things 'right'. At other times, we might do everything 'right' without getting the desired biodiversity impacts. We need to constantly evaluate, improve, and try out different things.

Conservationists are under considerable scrutiny. Many debates have taken place on the successes and failures of community-based conservation. We find ourselves under pressure to demonstrate that we are 'succeeding'. Similarly, and unfortunately, we are also under pressure to underplay our seeming 'failures'.

Implicit in the notion of success and failure, however, is a sense of closure, which, as I have mentioned before, does not have a place in conservation, and especially not in community-based efforts. Lack of progress must not make us lose faith in community-based conservation. After all, despite global and national efforts, tigers *Panthera tigris* have gone extinct inside Protected Areas, mining and large scale destruction have taken place inside National Parks, and wildlife reserves continue to be denotified under pressures of economic development. But that hasn't made us lose faith in the value of Protected Areas. Why should a few setbacks make us lose faith in the value of community-based conservation?

If we don't get the desired results, it is not a failure of community-based conservation. Instead, we need to learn from experiences and improve our interventions and partnerships with local communities; do things differently, or do different things.

Similarly, when we do get the desired results, it surely is a cause for celebration. However, claiming them to be 'successes' might be overblown, because they are temporary. The socio-ecological systems we work in and the threats to biodiversity are too dynamic to allow for any sense of closure or complacency.

Revisiting ownership: the right to conserve

A few years ago, I was interacting with Garma, a community leader on the Tibetan Plateau. Garma has played a key role in leading community-based conservation efforts. I was relying on interpreters, as I did not understand Tibetan, Garma's language. Even though I didn't understand his words, I could feel the frustration in Garma's voice as he explained how keen their community was to apprehend illegal poachers and miners on their land, but that they did not have the right to do so.

Some two years later, traveling in the Kyrgyz Tien Shan, I heard exactly the same words, albeit in a different language. This time they came from Kenesh, a fellow community leader from one of our partner communities in this area. He too expressed the community's helplessness in apprehending outside poachers.

Both Garma and Kenesh, in their respective native wisdom, had underscored a rather important internal disconnect between the philosophy and practice of community-based conservation. Community-based efforts tend to try and extend the responsibility of conservation to local communities, without always extending them the authority for it.

Similarly, in the predominant governance model for conservation in most parts of the world, wildlife is considered as belonging to the state. Community-based conservation, on the other hand, tries to promote local ownership over natural resources and wildlife, and it seems to do so without a legal basis.

So what is the way ahead? I don't pretend to know the answer. However, it is clear that for this to change would require immense courage on part of government leaders. To accord authority and ownership to local communities while maintaining an overseeing role would require a fundamental change in our historical hierarchical mindsets. Governments, just like individuals, find it difficult to devolve authority.

Hope and appeal for community-based conservation

Sensible and foresighted governance will be necessary for economic development to take place without incurring the immense cost to biodiversity and the environment that is all too common today. Conservationists therefore must work much more closely with governments.

It is useful to once again recognize here that, just like local communities, government is not a homogeneous entity either. There are differences in the mandates of the various Government departments that have implications for on-ground conservation (see Chapter 9: STRATEGIC SUPPORT). When it comes to conservation and communities, at an even more fundamental level, there tends to be a disconnect between two of the three most important arms of any government, the legislature and the executive.

Wildlife managers represent the executive or the bureaucracy. Much like conservationists, they have generally tended to view local communities as the source of conservation problems; an unwarranted player in power equations; and an undesirable reality in the wilderness that needs to be managed. At the extreme, this gives rise to the view of the community as the perpetrator. Indeed, there is abundant research in literature, including our own (e.g. Mishra et al. 2001, 2004, 2010), to show that resource use by local communities is an important cause of wildlife degradation.

On the other hand, for the politician, who constitutes the legislature, it is usually not the local people, but the conservation agenda that is an irritant that comes in the way of development and resources. The history of coercive conservation, the historical curtailment of resource access for local communities, and the damage caused by wildlife, tend to give local legitimacy to this political position. In addition, in electoral politics, long-term consequences notwithstanding, it seems helpful to have a common enemy that can unite the voters. Conservation can be construed to be the enemy against which to unite and fight.

In many situations, community-based conservation, however, may have a rather willing ally in the politician. Its focus on empowerment and welfare of local communities to achieve conservation tends to often – though not always – align well with the priorities of the political class. This is in complete contrast to how the local politician has traditionally tended to view conservation.

Community-based conservation, thus, may already have a potential ally in the legislature – and that can be better used to strengthen conservation laws, policies and systems. Capitalizing on this rare opportunity is essential, as even the future of Protected Areas – generally considered to be the last hope for biodiversity conservation – stands uncertain today. Community-based conservation gives hope not just for making Protected Areas more resilient, but also for extending the scope of conservation well beyond their boundaries.

We live in an increasingly democratic world. If we believe in the true spirit of liberal democracy, it shouldn't be hard to recognize that unless the majority of the population supports conservation, especially the local communities, we will continue to lose our biodiversity.

The main determinant of our ability to save the planet's biodiversity into posterity will not be the size or other metrics of Protected Areas, but our fundamental approach towards people in conservation.

References

Abubakar, M., Rajput, Z.I., Arshed, M.I., Sarwar, G., and Ali, Q. 2011. Evidence of peste des petits ruminants virus (PPRV) infection in Sindh Ibex (*Capra aegagrus blythi*) in Pakistan as confirmed by detection of antigen and antibody. Trop. Anim. Health. Prod. 43: 745-747

Bagchi, S., Mishra, C. and Bhatnagar, Y.V. 2004. Conflicts between traditional pastoralism and conservation of Himalayan ibex (*Capra sibirica*) in the Trans-Himalayan mountains. Animal Conservation 7: 121-128

Bangs, E.E., Fontaine, J.A., Jimenex, M.D., Meier, T.J., Bradley, E.H., Niemeyer, C.C., Smith, D.W., Mack, C.M., Asher, V., and Oakleaf, J.K. 2005. Managing wolf-human conflict in the north western United States. People and Wildlife: Conflict or Coexistence? (eds R. Woodroffe, S. Thirgood & A. Rabinowitz), pp. 340-356. Cambridge University Press, New York.

Berkes, F. 2007. Community-based conservation in a globalized world. PNAS 104: 15188-15193

Bhagwat and Rutte 2006. Sacred groves: potential for biodiversity management. Frontiers in Ecology and the Environment 4: 519-524

Chundawat R.S., Rawat G.S. 1994. Indian cold deserts: a status report on biodiversity. Wildlife Institute of India, Dehradun, India.

Cohen, J.R. 2001. When people are the means: negotiating with respect. Geo. J. Legal Ethics 14: 739-802

Coleman, J.S. 1986. Social Capital in the Creation of Human Capital. The American Journal of Sociology, 94, Supplement: Organizations and Institutions: Sociological and Economic Approaches to the Analysis of Social Structure pp. S95-S120

Colman, A.M. 1999. Game theory and its applications in the biological and social sciences. International series in Social Psychology. Routledge, London, UK.

Dagleish, M.P., Ali, Q., Powell, R.K., Butz, D. and Woodford, M.H. 2007. Fatal Sarcoptes scabiei infection of blue sheep (Pseudois nayaur) in Pakistan. Journal of Wildlife Disease 43: 512-517

Dietz, T., Ostrom, E., and Stern, P. C. 2003. The struggle to govern the commons. Science, 302(5652): 1907-1912

Dudley, N. (Editor). 2008. Guidelines for Applying Protected Area Management Categories. Gland, Switzerland: IUCN. x + 86pp.

Fisher, R., Ury, W., Patton, B. 1991. Getting to yes. Negotiating an agreement without giving in. 2nd ed. London: Random House Business Books.

Fox, J.L., Nurbu, C., Bhatt, S. and Chandola, A., 1994. Wildlife conservation and land-use changes in the Transhimalayan region of Ladakh, India. Mountain Research and Development 14:39-60.

Gartzke, E., Li, Q., and Boehmer, C. 2001. Investing in the peace: economic interdependence and international conflict. International Organization 55: 391-438

Gerdes, K.E., and Segal, E. 2011. Importance of empathy for social work practice: integrating new science. Social Work 56: 141-148.

Gambrill, E. 2012. Social work practice: a critical thinker's guide. Oxford University Press. New York.

Goldstein, M. C. and Beall, C. M. 1989. Nomads of western Tibet: survival of a way of life. Serindia Publications, London. 191 pp.

Graham-Rowe, D. 2011. Endangered and in demand. Nature 480: S101-S103.

Gusset, M., Swarner, M., Mponwane, L., Keletile, K. & McNutt, J. 2009. Human-wildlife conflict in northern Botswana: livestock predation by endangered African wild dog *Lycaon pictus* and other carnivores. Oryx 43: 67-72

Hardin, G., 1968. The tragedy of the commons. Science 162: 1243-1248

Jackson, R. and Wangchuk, R. 2001. Linking snow leopard conservation and people-wildlife conflict resolution: grassroots measures to protect the endangered snow leopard from herder retribution. Endangered Species Update 18: 141

Johansson, O., McCarthy, T., Samelius, G., Andren, H., Tumursukh, L., Mishra, C. 2015. Snow leopard predation in a livestock dominated landscape in Mongolia. Biological Conservation 184: 251-258

Karp, D.G. 1996. Values and their effect on pro-environmental behavior. Environment and Behavior 28: 111-133

Lamarque, F., Anderson, J. Anderson, Chardonnet, P., Fergusson, R., Lagrange, M., Osei-Owusu, Y., Bakker, L., Belemsobgo, U., Beytell, B., Boulet, H., Soto, B., and TabiTako-Eta, P. 2008. Human-wildlife conflict in Africa: An overview of causes, consequencesand management strategies. Working Paper –ILO.

Ludwig, D., Mangel, M. and Haddad, B., 2001. Ecology, conservation, and public policy. Annual Review of Ecology and Systematics, 32: 481-517.

Madhusudan, M.D. 2003. Living with large wildlife: livestock and crop depredation by large mammals in the interior villages of Bhadra Tiger Reserve, southern India. Environmental Management 31: 466-475

Madhusudan, M. D. & Mishra, C. 2003. Why big, fierce animals are threatened:

conserving large mammals in densely populated landscapes. Pages 31-55 in: Saberwal, V. K. and Rangarajan, M. (eds.) Battles over nature: science and the politics of wildlife conservation. Permanent Black, Delhi

Maclennan, S. D., Groom, R. J., Macdonald, D. W. and Frank, L. G. 2009. Evaluation of a compensation scheme to bring about pastoralist tolerance of lions. Biological Conservation 142: 2419-2427

McCann K.S. 2000. The diversity-stability debate. Nature 405:228–33

Mishra, C. 1997. Livestock depredation by large carnivores in the Indian trans-Himalaya: conflict perceptions and conservation prospects. Environmental Conservation 24: 338-343

Mishra, C. 2000. Socio-economic transition and wildlife conservation in the Indian Trans-Himalaya. Journal of the Bombay Natural History Society, 97: 25-32

Mishra, C. 2001. High altitude survival: conflicts between pastoralism and wildlife in the Trans-Himalaya. Ph. D. Thesis, Wageningen University, The Netherlands

Mishra, C., Prins, H. H. T. and Van Wieren, S. E. 2001. Overstocking in the Trans-Himalayan rangelands of India. Environmental Conservation28(03) 279-283

Mishra, C., Allen, P., McCarthy, T., Madhusudan, M. D., Bayarjargal, A., and Prins, H. H. T. 2003a. The role of incentive programs in conserving the snow leopard. Conservation Biology 17 1512-1520

Mishra, C., Van Wieren, S. E., and Prins, H. H. T. 2003b. Diversity, risk mediation, and change in a Trans-Himalayan agropastoral system. Human Ecology 31: 595-609

Mishra, C. and Fitzherbert, A. 2004. War and wildlife: a post-conflict assessment of Afghanistan's Wakhan Corridor. Oryx 38:102-105

Mishra, C., van Wieren, S.E., Ketner, P., Heitkonig, I.M.A. and Prins, H.H.T. 2004. Competition between livestock and bharal Pseudois nayaur in the Indian Trans-Himalaya. Journal of Applied Ecology 41: 344-354

Mishra, C. and Suryawanshi, KR. 2014. Managing conflicts over livestock depredation by large carnivores. Pages 27-47 in: Successful management strategies and practice in human-wildlife conflict in the mountains of SAARC Region. SAARC Forestry Centre, Thimphu, Bhutan.

Mishra, C. and Suryawanshi, K.R.2015. Conflicts over snow leopard conservation and livestock production. In: Conflicts in Conservation: Navigating Towards Solutions, 16. Redpath S, Young J, Gutierrez R, Wood K (eds.), Cambridge University Press.

Mishra, C., Bhatnagar, Y. V., et al. 2016a. The role of village reserves in revitalising the natural prey base of the snow leopard. Pp.184-195 in: Snow Leopards. McCarthy, T., and Mallon, D. (eds.), Elsevier.

Mishra, C., Redpath, S.R. and Suryawanshi, KS. 2016b. Livestock predation by snow leopards: conflicts and the search for solutions. In McCarthy, T. and Mallon, D. (eds.). Snow leopards of the world. Elsevier. In press.

Morrison, K., Victurine, R., and Mishra, C. 2009. Lessons learned, opportunities and innovations in human wildlife conflict compensation and insurance schemes. WCS Translinks Program.

Muruthi, P. 2005. Human wildlife conflict: lessons learned from AWF's Africa heartlands. African Wildlife Foundation. Working Papers.

Namgail, T., Bhatnagar, Y.V. and Fox, J.L. 2007. Carnivore-caused livestock mortality in Trans-Himalaya. Environmental Management 39: 490-496

Naughton-Treves, L., Grossberg, R., and Treves, A. 2003. Paying for tolerance: Rural citizens' attitudes toward wolf depredation and compensation. Conservation Biology 17: 1500-1511

Nelson, M. P., Bruskotter, J. T., Vucetich, J. A., & Chapron, G. (2016). Emotions and the ethics of consequence in conservation decisions: Lessons from Cecil the Lion. Conservation Letters. doi: 10.1111/conl.12232

Nyhus, P., Fischer, H., Madden, F. and Osofsky, S. 2003. Taking the bite out of wildlife damage: the challenges of wildlife compensation schemes. Conservation in Practice 4: 37-40

Oli, M.K., Taylor, I.R. & Rogers, M.E. 1994. Snow leopard Panthera uncia predation of livestock: an assessment of local perceptions in the Annapurna conservation area, Nepal. Biological Conservation 68: 63-68

Olson, M. 2009. The logic of collective action (Vol. 124). Harvard University Press.

Ostrom, E. 1990. Governing the commons: The evolution of institutions for collective action. Cambridge University Press.

Ostrowski, S., Thiaucourt, F., Amirbekov, M., Mahmadshoev, A., Manso-Silván, L., Dupuy, V., Vahobov, D., Ziyoev, O., and Michel, S. 2011. Fatal outbreak of Mycoplasma capricolum pneumonia in endangered markhors. Emerging Infectious Diseases 17: 2338-2341

Portes, A. 1998. Social capital: its origins and applications in modern sociology. Annual Review of Sociology 24: 1-24

Redpath, S. M., Young, J., Evely, A., Adams, W.M., Sutherland, W.J., Whitehouse, A., Amar, A., Lambert, R.A., Linnell, J.D., Watt, A. and Gutiérrez, R.J., 2013. Understanding and managing conservation conflicts. Trends in Ecology &

Evolution, 28: 100-109. doi: 10.1016/J.TREE.2012.08.021

Rittel H, Webber M. 1973. Dilemmas in a general theory of planning. Policy Sciences 4:155–69

Salafsky, N. and Margoluis, R. 1999. Threat reduction assessment: a practical and cost-effective approach to evaluating conservation and development projects. Conservation Biology 13,830-841

Schwartz, S.H. 2006. A Theory of cultural value orientations: explication and applications. Comparative Sociology 5: 137-182

Smutko, L.S. 2005. Negotiation and collaborative problem solving. Natural Resources Leadership Institute, NC State University.

Suryawanshi, K. R. 2013. Human carnivore conflicts: understanding predation ecology and livestock damage by snow leopards. Ph.D. Thesis. Nature Conservation Foundation and Manipal University

Suryawanshi, K.R., Bhatnagar, Y.V., Redpath, S.R., and Mishra, C. 2013. People, predators and perceptions: patterns of livestock depredation by snow leopards and wolves. Journal of Applied Ecology. doi: 10.1111/1365-2664.12061.

Suryawanshi, K.S., Bhatia, S., Bhatnagar, Y.V., and Mishra, C. 2014. Multi-scale factors influencing human attitudes towards snow leopards and wolves. Conservation Biology 28: 1657 - 1666. doi: 10.1111/cobi.12320

Treves, A., Jurewicz, R. L., Naughton-Treves, L., and Wilcove, D. S. 2009. The price of tolerance : wolf damage payments after recovery. Biodiversity & Conservation 18: 4003-4021. doi: 10.1007/s10531-009-9695-2

Index